Ethics for people who work in tech

This book is for people who work in the tech industry—computer and data scientists, software developers and engineers, designers, and people in business, marketing or management roles. It is also for people who are involved in the procurement and deployment of advanced applications, algorithms, and AI systems, and in policy making. Together, they create the digital products, services, and systems that shape our societies and daily lives. The book's aim is to empower people to take responsibility, to 'upgrade' their skills for ethical reflection, inquiry, and deliberation. It introduces ethics in an accessible manner with practical examples, outlines of different ethical traditions, and practice-oriented methods. Additional online resources are available at: ethicsforpeoplewhoworkintech.com.

Ethics for people who work in tech

Marc Steen

CRC Press

Taylor & Francis Group
Boca Raton London New York

CRC Press is an imprint of the
Taylor & Francis Group, an **informa** business

A CHAPMAN & HALL BOOK

Cover Image Credit: Elodie Oudot on Unsplash

First Edition published 2023
by CRC Press
6000 Broken Sound Parkway NW, Suite 300, Boca Raton, FL 33487-2742

and by CRC Press
4 Park Square, Milton Park, Abingdon, Oxon, OX14 4RN

CRC Press is an imprint of Taylor & Francis Group, LLC

ISBN: 978-0-367-54330-3 (hbk)
ISBN: 978-0-367-54243-6 (pbk)
ISBN: 978-1-003-08877-6 (ebk)

DOI: 10.1201/9781003088776

Typeset in Sabon Lt Std
by KnowledgeWorks Global Ltd.

To my mother, Nan, and my late father, Bert,
who taught by example, about ecological
and social justice, and about learning and
sharing knowledge.

Contents

Part I

Motivation and key concepts

In this part (Chapters 1–7), we explore *why* you would want to integrate ethics in your work, in your projects. We explore a view on ethics that understands ethics as a process; as *doing ethics*, as efforts to promote ethical reflection, inquiry, and deliberation. We explore how we can move our scope by zooming-in and zooming-out, and how we can move back and forth between practice and theory. Then we discuss ethics as a domain of knowledge, in relation to other domains of knowledge. We will discover similarities between normative ethics and the design and application of technology. We will also discuss the notion that technologies are *not* neutral tools. Then we briefly discuss two topics that are relevant for our discussion of technology and ethics: wellbeing and economics. These are two ways to talk about creating value, and they often provide a background to projects. Technologies are designed and deployed in order to promote people's wellbeing and/or to create economic value. We also look at three topics that regularly feature in discussions of technology and ethics: the 'trolley problem', privacy, and responsibility.

DOI: 10.1201/9781003088776-1

Chapter 1

A humanistic approach

Technology has become increasingly important in our societies and in our daily lives. Some see technology as a force for good, at least potentially. And indeed, computers, algorithms, robots, and artificial intelligence (AI) systems can be used for beneficial purposes. For example, to enable sustainable energy and mobility, or to improve healthcare and education. Others draw attention to the negative impacts of technologies on society, how their applications can corrode or undermine values like equality, freedom, privacy, solidarity, and democracy; and can have negative impacts on our daily lives and on our ways of relating and interacting with each other.

DOI: 10.1201/9781003088776-2

We have seen the role of social media in spreading fake news, stirring-up polarization, and disturbing elections. We may fear the ways in which algorithms, robots, and AI systems can disrupt work, employment, and daily life. We have also seen attempts to rein in and regulate technology. In 2016, for example, the European Commission's *High-Level Expert Group on AI* published *Ethics Guidelines for Trustworthy AI*, with recommendations for a human-centric approach to the design and application of AI systems that are lawful, ethical, and robust. They advocate principles like respect for human autonomy, prevention of harm, fairness, and explicability. In 2019, the *IEEE*, a large international professional organization, published *Ethically Aligned Design*, in which they outline '*A Vision for Prioritizing Human Well-being with Autonomous and Intelligent Systems*' (the publication's subtitle). We can understand efforts like these as being part of a *humanistic* approach to technology; an approach that takes human dignity and human autonomy as starting points for the design and application of technology.

I can see both positive and negative aspects of tech. And I tend to lean towards critique more than towards praise, to counterbalance the sometimes excessive, uncritical enthusiasm for tech.

I would be curious where you stand; how you think and feel about tech. You picked up this book and started reading. My guess is that you work, or want to work, in the technology sector, in computer or data science, in software development or engineering, in design, marketing or management, or in some other role. I also guess that you are interested in ethical and societal issues regarding the design and application of technologies and that you are motivated to do good. Please note that, throughout this book, despite my critical tone in some places, I will never mean to judge anybody. I assume that all of us can have positive intentions, that we all have our particular inclinations and challenges, and that, in various ways, we all try our best.

Bit about my background

This book is the product of over 25 years of working in various organizations, in diverse roles: research, design, consultancy, and project management. I earned MSc, PDEng (professional doctorate in engineering), and PhD degrees at the Faculty of Industrial Design Engineering of Delft University of Technology in the Netherlands. Its pioneering curriculum in human-centred design, transdisciplinary innovation, and innovation management has shaped my outlook on technology and innovation: to put people and their experiences centre stage, to seek collaborations across disciplinary boundaries, and to focus on facilitating the innovation process. The Delft curriculum has informed many other schools, across the globe.

After working at Philips Electronics' Multimedia Lab and KPN Research (at the time the Research & Development lab of the Dutch incumbent telecom operator), I joined TNO, The Netherlands Organization for applied scientific research, a research and technology organization with some 3600 employees, where I have continued to work until the present.

Over the years, I worked on diverse projects, for example: introducing a platform for online education in primary schools (1997); creating an online community for people who suffer from rheumatism (1999); experiments with designing an *Electronic Program Guide* for TV, with personalized recommendations (2000); business models for collaboration between parties in the design and deployment of mobile internet services, one case with police officers, and one case with people who provide informal care to people with dementia (2000–2008), designing *I-mode* services (an early form of mobile internet, developed by Japanese NTT DoCoMo, introduced in Europe in 2005); user experience experiments with intellectual property rights protection for buying music on *I-mode*; experiments with designing mobile phones that elicit 'wow' experiences (2003–2005); exploring future needs and services in telecom, in co-creation with children (2007); prototypes of telecom and multimedia services to promote togetherness, in an international research consortium, with British Telecom and others (2008–2012); creating online services to support people who provide informal care to people with dementia, also in an international consortium (2010–2012); an international project to promote open innovation in the development and adoption of LED technologies (2013–2016), facilitating collaboration between academia and industry, and with suppliers and customers; a project with cars and sensors, which aimed to reduce traffic and promote public transport (2016); projects to promote collaboration between police officers and citizens in co-creating public safety, for example, through online 'neighbourhood watch' groups; research into fairness and transparency in the design and deployment of algorithms in the domain of justice and security (2018–2020); and creating workshop formats to integrate ethical, legal, and societal aspects (ELSA) in the design and deployment of specific innovations, notably in the domain of justice and security.

Over the years, societal engagement, the involvement of prospective users, and promoting ethical deliberation processes have been recurring themes. Between 2004 and 2008, these interests led me to participate in a part-time PhD program at the University of Humanistic Studies and to write a thesis that I defended in Delft. Over the years, I continued to read books and articles and engaged in research, which led to numerous articles on *Human-Centred Design*, *Value Sensitive Design*, and *Responsible Innovation*; both academic and popular, and often in-between, to connect theory and practice.

My purpose is to share knowledge that I have collected over the years, to help you 'upgrade' your moral sensitivities and your skills for ethical reflection, inquiry, and deliberation, so that you can do the right thing in your work and in your projects. My hope is that you can help, with your work, to create conditions in which people can flourish. Or more precisely: to create conditions in which diverse people are enabled to live their versions of *the good life*. Because people differ.

Ethics in and between people

I understand ethics and responsibility as happening within people and between people, and I will focus on ethics and responsibility as they happen in the diverse phases of innovation: design, development, application, deployment, and usage. Consequently, I will *not* talk about 'ethical products' or 'responsible systems'. I find those terms problematic. In my understanding, *people* can behave ethically or responsibly. Products or systems cannot; they are machines. Much of my work falls under the broad term *Responsible Innovation*, which refers to efforts to align innovation, including the design and application of technologies, with needs and concerns in society. Here, *responsible* refers to the innovation *process*, to people's efforts to act responsibly.

Because of my focus on the ethics that happen in and between people, I will say little about attempts to program ethics into machines. There are excellent books on that topic, on *machine ethics*; for example, *Moral Machines*, by Wallach and Allen. They discuss tensions between, on the one hand, promoting *human* autonomy and, on the other hand, promoting autonomy in *systems*; between human control and autonomous systems.

This distinction, however, between ethics in *people* and ethics in *machines*, is not always clear.

Imagine professionals who have the task to detect fraudulent or criminal behaviour, for example, in an insurance company, government agency, or police organization. They use a decision support system that places orange flags behind the names of persons for which an algorithm calculated a high risk of fraudulent or criminal behaviour. These calculations are typically based on data from the past. Imagine also that these professionals are expected to not merely follow the algorithm's output, but to also use their professional, discretionary competence. They can choose to follow the system's orange flags, or provide push-back, to modify and correct its outputs. They have moral agency. But using the system can influence their moral agency. Over time, for example, using the system can steer their decisions. This may happen implicitly and unintentionally, as if they slide on a slippery slope, approving the system's outputs increasingly unthinkingly. Meanwhile, they remain accountable. They need to be prepared to answer questions from their manager, from an auditor, or from a journalist. They cannot point at the computer and say: 'It did it'. In such a situation, moral agency is sort of distributed: between the people and the decision support system, and probably also between different people in the organization. In that case, we have the 'problem of many hands'. This is not a new problem; it has existed since people have collaborated and have shared tasks and responsibilities. But it can become rather complex when machines become part of the decision-making process, when they add yet more 'hands'.

In sum, my view on ethics as happening in and between people, and not in machines, is not as simple as it would look at first sight; it does take into account interactions with machines.

A fuller discussion of these issues would need to go into *Human-Machine Teaming*, which deals with organizing collaboration between people and machines, and into *Meaningful Human Control*, which deals with enabling people to exercise control over the systems they use. I chose, however, to put these topics out of scope. You can, though, keep them in mind while reading.

Ethics as a process

Now, the topic of *ethics*. Some people associate ethics with assertions about good or evil, with finding definitive answers to moral questions, or with setting norms to limit people's behaviours.

I understand ethics differently: as a process, as a verb, as doing ethics. Ideally, we organize collaborative and iterative processes that make room for ethical reflection, inquiry, and deliberation, and that involve thinking, feeling, and action: your head, your heart, and your hands.

Moreover, throughout this book, I will discuss different normative orientations, in order to support you in reflection, inquiry, and deliberation, so that you can help to steer projects in directions that are (more) desirable; in directions that promote justice and wellbeing.

Motion and process are recurring themes in this book. They play in how I understand the innovation process: as a flow that starts with people and ideas, then moves to defining and organizing projects, then the design and application of specific products or systems, which then flow into the world, into society and social systems, and into people's daily lives and habits.

Motion is also at play in my invitations to you to zoom-out and to zoom-in. At times, I will invite you to zoom-out and see the whole. At other times, I will invite you to zoom-in and see the parts. Zooming-out can help to investigate and discuss patterns in society; for example, the ways in which the project you work in can help or hinder in shaping society in a specific direction, for example, regarding social or economic (in)equality. Conversely, zooming-in can help to discuss specific elements of a project; for example, the business model that it is premised upon, or the ways in which a blinking or beeping user interface feature can nudge people into behaviours that serve this specific business model.

In my experience, the whole and the parts can become disconnected disadvantageously. Typically, on the level of an organization's or a project's leadership, people talk about high-level goals, like 'build community' or 'foster collaboration', whereas, on the shop floor or in projects, people focus on practical goals, like 'increase people's usage of this app' or 'get the system running'. One of my hopes is that we can learn to better connect these two levels, so that leadership and projects can learn from and support each other.

Innovation process

Now is a good moment to speak a bit about innovation. I understand innovation as a *process* that involves people and things: researchers, developers, customers, and users; and objects, materials, technologies, and machines. It involves both 'design' and 'usage'; between inverted commas because, especially with networked and interactive media, the line between design and usage is blurred. Users can behave creatively and innovate in how they use tech. And design often depends on people using the tech in particular ways. Furthermore, innovation involves economics and money, and politics and power. In short, innovation is a complex process. It has multiple inputs: the knowledge, ambitions, and efforts of the people involved; and multiple outputs: new practices that are enabled by the innovation (here understood as a *noun*), and the wider impacts it has on people's daily lives and on society. The people involved may be experts, but their knowledge of and control over what happens in the innovation process is limited. Therefore, I would be wise if we are aware of these limits, and act carefully and with humility.

Crucially, I would like to clarify that innovation does not necessarily centre around technology. Sure, there are plentiful examples of technological innovations. But there are also many examples of social innovation; for example, of new ways to organize work, new behaviours and habits in daily life, and new patterns and distributions of power in society. Let me give some more examples of social innovation. You may think of them as alternatives to technological innovation. Instead of working on an innovative vehicle, you can think about alternative ways to fulfil the need to travel or the desire to meet people and socialize. You can think of improving the quality and accessibility of public transport, of busses and trains, rather than create yet another vehicle for only one or several people. That would probably also save material and energy. Or, regarding the desire to meet new people, possibly of other cultures, you can think of organizing joint dinners, in your local city, with people from other cultures, maybe ex-pats, maybe immigrants, with all sorts of different cuisines. Including grocery shopping at a local market and cooking together. You do not have to travel abroad to have adventures. You can also think of new services that are social in nature, rather than based on technology. Imagine that you work for a company that offers telecommunication products and services. Instead of focusing on more bandwidth, less delay, better audio and more pixels, they could partner with a company that offers training of communication skills and jointly offer courses in non-violent communication. Because, well, communication is not about network connection, but about human connection. Or, if your assignment is to help promote social cohesion in a particular city, you do not have to think immediately of online tools; rather, you may want to explore ways in which citizens can meet in a physical space and facilitate communication between them, maybe a face to face event with follow-up actions online.

Systems thinking

This need for combining zooming-out and zooming-in borrows from *systems thinking*, which deals with seeing both the parts and the whole; both the trees and the forest.

Figuratively, it advocates looking at the trees, their branches, their leaves and flowers, and at the insects that help to spread their pollens so that fruits can develop, at the soil and its microorganisms and fungi that are vital to the trees' nutrition, and at the smaller and larger animals and birds that live in the forest, and, importantly, at the intricate relationships between all these organisms. The field of systems thinking was pioneered by Donella Meadows, lead author of *Limits to Growth* (1972), which put sustainable development on the agenda of governments and businesses.

Systems thinking typically involves making a model of a complex reality. Imagine a sheet of paper with circles drawn on it. Each circle represents a phenomenon; for example, car movements, fuel price, public transport, and bike movements. In a real analysis, there will be many more elements. And there are arrows between circles, to represent relationships. For example, higher fuel prices can reduce car movements, maybe only to some extent or temporarily. And the availability and quality of public transport can lessen car movements, for some types of travel. A real analysis will, of course, be much more complex and elaborate.

A key concept in systems thinking (and in cybernetics, a precursor of the field of AI) is the *feedback loop*; it models a particular type of interaction between different elements of a system. Say that A influences B, then a feedback loop will, over the course of time, bring information about the status of B back to A and will regulate A and its influence on B. Critically, a feedback loop can be either balancing or reinforcing. A balancing feedback loop brings a dynamic equilibrium to a part of the system. When people experience that motorways are busy, they can replace the car by public transport or bike. A reinforcing feedback loop makes parts of a system go increasingly up or increasingly down. In response to busy motorways, public administrators, possibly under pressure of lobbying parties, can choose to build or widen motorways, which can result in more car movements, and, over time, even busier motorways, and a perceived demand for even more motorways.

In systems thinking, it is critical to look beyond the phenomena that are visible on the surface and to look at underlying relationships and interactions. For the design and application of algorithms, for example, this would involve looking beyond the algorithm in a narrow sense. It would involve looking at the process and the organization in which the algorithm will be used, how it will be used, and how its practical usage may affect processes over the course of time. For example, the positive or negative impacts on the ways in which employees can collaborate, or on the interactions between employees and customers, or public agents and citizens. Moreover, we need to look at feedback loops, which we can understand as vicious circles or virtuous circles; do they help to create desirable outcomes or do they lead to undesirable outcomes?

Practice and theory

Another recurring theme is the move back and forth between theory and practice. We will look at theoretical concepts and at practical examples. At times, I will invite you to reflect on your own practices. This way of combining practice and theory borrows from pragmatist philosopher John Dewey (1859–1952). Dewey found the distinction between practice and theory unproductive. Instead, he advocated combining practice and theory: to start any project by engaging with people's practical experiences and problems; then move to reflection and theory in order to generate knowledge; and then move back to people's practices and apply that knowledge to help solve their practical problems. We will further explore these themes when we discuss: the need for an iterative approach in problem-setting and solution-finding (in Chapter 17 on Human-Centred Design); ways to invite actors and stakeholders to express their values (in Chapter 18 on Value Sensitive Design); and principles like anticipation, responsiveness, inclusion, and reflexivity (in Chapter 19 on Responsible Innovation).

Stop, think, act

Years ago, when my girlfriend Cynthia and I were on holidays in Milos, Greece, we met Iannis. He was our dive instructor and he taught us that when there is a problem, you need to do the following: *stop, think, act*. Stop what you are doing, so that you can observe more clearly what is going on. Then think about the problem, its possible causes, explore potential solutions, and communicate with your dive buddy and others. And then act: effectively and efficiently.

Action without thinking is clueless. Thinking without action is useless.

What does that mean for a development or implementation project? It is not very common that someone says 'stop' during a project. Typically, the project's scope and goal are negotiated and agreed upon, and deliverables and deadlines are written down in plans. Raising an issue that is within scope and that is relevant for a deliverable or a deadline is, of course, welcomed. Saying something like 'In our project, I find X problematic' is, however, often less welcome.

Yet, reading this book can make you question views that are taken for granted, or express concerns that are not on the agenda. For me, *doing ethics* entails asking questions that can be uneasy, creating situations that can feel awkward, and tolerating moments of tension and uncertainty.

'The unexamined project is not worth working on', to paraphrase Socrates.

The process of learning to scuba dive can also illustrate the process of expanding your awareness and abilities in the domain of ethical reflection, investigation, and deliberation. In my very first dives, I focused on using the equipment: to breathe through the regulator in my mouth and to control buoyancy by inflating and deflating my vest. I was unable to see much beyond my instructor and my dive buddy. In later dives, I learned to breathe and float with less effort. Then I could see my surroundings: fishes, corals, snails, plants, and rocks. Still later, I learned to use a compass and to navigate the underwater terrain and to plan and execute safe and pleasant dives. A next step, for me, could be to learn to organize dives for a group of, say, three other divers, as a dive guide.

Gradually, my circle of awareness and my abilities to exercise control expanded. I think a similar process is possible for your abilities for ethical reflection, inquiry, and deliberation. You are invited to first dip a foot in the water, then gradually immerse yourself in theories and concepts, and expand your ethical sensitivity and dexterity. You may want to try out methods for ethical deliberation in your projects and plan workshops that are safe and productive. Maybe, over time, you will want to learn how to take others with you. Like scuba diving, *doing ethics* can be an adventure!

Throughout this book, I will mention books, reports, and articles that I have found inspiring. You can read a bit more about these in the Notes and further reading at the end of the book. I hope that you will read some of these. Before diving in, you may prefer to view online talks or presentations by their authors, or listen to interviews or podcasts with them. For me, this has often been a nice way to get acquainted with new perspectives and ideas.

Throughout this book, you will be invited to reflect on your work and projects. To start off, could you take one project that you currently work on, or did work on, or will work on? Maybe in a consultancy role, a management role, or a supporting role? Are there issues in this project that you experience as 'ethical' or problematic? Which issue in particular? What do you think about this issue? How do you feel about it? Can you zoom-out to see a larger whole? Can you zoom-in to take a closer look? How would you like to act? Any impulse that you can feel in your body?

Chapter 2

What do we mean with *ethics*?

I have heard people say that 'ethics is not a real science' or that 'ethics is only an opinion'. Maybe you have heard similar things. Or maybe you have thought of them. In the current chapter, I will discuss ethics as a domain of knowledge and its relation to other domains of knowledge. This may feel like a bit of a detour, but I hope that you can stay with it and that it will pay off.

Different domains of knowledge

I would like to invite you to imagine a tree. It is a large tree, so you may need to take a couple of steps back to see it properly. This tree represents scientific knowledge. You see four main

DOI: 10.1201/9781003088776-3

branches: the natural sciences, the social sciences, the humanities, and technology. People in the *natural* sciences study nature in fields like physics, chemistry, earth sciences, and biology. In the *social* sciences, people study human behaviour and social structures, in fields like psychology, anthropology, sociology, economics, and organization studies. People in the natural and social sciences typically generate knowledge through observation and experimentation.

People in the *humanities* study the products of people's minds and cultures in fields like history, law, language and literature, and philosophy. They may conduct empirical studies, possibly together with people in the natural or social sciences, or they may generate knowledge in ways that are typical for the humanities: by reading, discussing, and writing books and articles.

The fourth branch represents *technology*, in which people combine, generate, and apply knowledge in order to produce means to achieve ends in fields like design, engineering, and computer science. In contrast to people in the natural and social sciences, who aim to describe and understand current realities, people in technology envision future realities and help to create these. They can use knowledge from the natural and social sciences and from the humanities.

Moreover, there are fields like mathematics or logics, which can be used in multiple domains.

If we look at the branch of the humanities and take a closer look at philosophy, we can see several offshoots, in which different types of questions are asked: *metaphysics*, with questions about reality and existence; *epistemology*, with questions about knowledge and truth; *aesthetics* with questions about art and beauty; and *ethics* or *moral philosophy*, with questions about morality and 'the good life'. Zooming-in further on ethics, we can see three twigs: *descriptive* ethics, which deals with describing people's beliefs about morality, values, and norms; *meta*-ethics, which studies ethics more theoretically; and *normative* ethics, which deals with practical questions about *how to live well* and *doing the right thing*. This book deals mainly with the latter, with *normative* ethics, which is related to other fields of ethics, like professional ethics, engineering ethics, business ethics, ethics of technology, and ethics of innovation.

Normative ethics and technology

The current exploration is meant to provide a background to present two ideas: that each of these different knowledge domains has its own particular assumptions, commitments, and methods; and that normative ethics and technology are more similar than you might think at first glance.

The natural and social sciences are concerned with describing phenomena in the world. In contrast, normative ethics and technology are concerned with changing things in the world. They are concerned with finding ways to deal with moral issues and to live well together, and envisioning and creating means to bring about positive changes in the world.

Moreover, normative ethics and technology share similarities. Both involve the following: looking at the world and experiencing something as problematic, finding that a situation is not quite right and needs to be changed for the better; and envisioning ways to improve that situation, for example, conceptually, by proposing perspectives to better understand the problem at hand, or practically, by creating tools that can help to solve that problem.

Of course, that does not mean that everything that technologists do is 'ethical'. Not in the way that 'ethical' is often understood: as praiseworthy, as the *right thing to do*. Imagine a company that is concerned primarily with making short-term financial profit. This will affect their policies and activities. A person who believes that companies need to steer their activities towards sustainable development (contribute to people, planet, prosperity) will probably find this company 'unethical'.

Realism and constructivism

Now, I do not suggest that normative ethics and technology are the same. I merely propose that they share similarities. Moreover, it is critical to acknowledge that different domains of knowledge have different assumptions, commitments, and methods. In the natural and social sciences, one can settle an argument by providing evidence that others find convincing. Please note that I said 'evidence that others find convincing' rather than 'evidence that is true'. With this, I take a constructivist stance, rather than a realist stance. Realists believe that reality exists independently of the researchers and the conceptual and experimental tools they use to study reality. In contrast, constructivists believe that researchers construct knowledge and that the tools they use can influence their findings. Constructivists are interested in the conceptual and experimental tools that researchers use, and in the assumptions that come with these tools and which affect how they look at the world. Which topics or issues become visible, and which topics or issues remain invisible, with this or that tool and assumption? When we use this or that tool, which types of data are included (or excluded) in a study, survey, or analysis?

Constructivists may use methods from sociology or history to study the processes via which people generate knowledge. French philosopher and sociologist Bruno Latour, a proponent of constructivism, studied researchers' activities in order to understand this construction process. In *Laboratory life: The social construction of scientific facts*, he studied the practices of a group of scientists at the *Salk Institute*, a prestigious biology lab, using the methods a sociologist would use to study a foreign tribe. And in *Aramis, or the Love of Technology*, he studied the efforts of engineers to create an innovative transport system in Paris, using the methods of an investigative journalist. His studies draw attention to the role of power, both economic and political, in the practical construction of scientific knowledge and of technological artefacts. Latour's findings put into question some common beliefs of how science works.

In a short essay, *Three Little Dinosaurs*, Latour talks about *Realsaurus*, *Scientosaurus*, and *Popsaurus*. The *Realsaurus* is the animal that lived in the Mesozoic Era. The *Scientosaurus* is the animal as it is known by palaeontologists. Strikingly, palaeontologists can update and modify their knowledge during scientific conferences: 'During three days of interminable discussion, the form of *Scientosaurus* had undergone a profound change. Before the meeting, he had been a cold-blooded lout, lazily slouching through in the swamps; by the end of the conference, he had warm blood, found himself very carefully streamlined, and ran about like a fine fellow trying out all sorts of new foods'. Each time that *Scientosaurus* changes, the *Realsaurus* changes also. Or does it? And the *Popsaurus*? That is the animal that most people envision if they think of a dinosaur; very likely the animal that they have seen in movies. Latour reminds us that we need to distinguish between *reality,* our *attempts* to investigate and understand reality, and people's *beliefs* about reality.

With this example, Latour wants to draw attention to the process of making scientific claims: when claims are accepted by relevant people, notably by fellow scientists, they are made true. (In practice, this makes more sense than the other way around: 'if facts are true, they are accepted'.) We cannot, however, take this to the extreme and say, for example, that gravity is a 'social construct'. Try make a brick float in mid-air. It will always fall, because, on earth, gravity is real. Moreover, in engineering, some facts are true because they work in practice. We know how to build a bridge that works with gravity, with forces of push and pull, with pull forces in steel cables and push forces in concrete blocks. The calculations we use have proven to be true because they work.

Relevance

Now, how is this relevant for your work? I guess that understanding the different worldviews that are at play in different knowledge domains and understanding the differences between realism and constructivism can help you to find ways to integrate ethics in your projects.

We can now appreciate that science and technology are partly 'hard' or 'objective', in the sense that a formula like $F = m \times a$ works well, regardless of economic or political power dynamics. And also that science and technology are partly 'soft' or 'subjective', in the sense that economic and political powers do shape the processes of research and innovation. Which projects do (not) receive funding? From which parties or agencies? What are their concerns and agendas? Which concepts and methods are you (not) allowed to use in your organization or in your project? Which values and interests are (implicitly) endorsed, with these concepts and methods? And which are (implicitly) ignored or stunted?

Whether you like it or not (or are aware of it or not), your work and your project are embedded in a network of actors, and each actor will try to excise power over your work and your project.

A helpful take on the role of different domains of knowledge comes from Danish economic geographer Bent Flyvbjerg. He observed that the social sciences often attempt to emulate natural science: to find 'objective' laws, make predictions, and construct experiments to produce 'hard' facts. Such attempts, however, often fail, for various reasons. In social science, people study people; researcher-people study subjects-people. Researchers, with particular values and interests, use methods and tools, which embody these values and interests, to study subjects, who also have values and interests. Inevitably, research and innovation happen in arenas of power differences and moral issues. Flyvbjerg proposed to see this not as a bug, but as a feature. He suggested that people in social science need to stop emulating natural science and instead engage in 'phronetic social science' (from *phronesis*, a term that Aristotle used to refer to practical wisdom): to make explicit the values and interests of the people and parties involved in a specific project. This would benefit both the social sciences and the natural sciences. People in natural science typically avoid making explicit the political and moral issues that are at play in their endeavours; they try to be 'objective'. This leaves a void, which people from social science can help to fill if they make explicit their values and interests, which they need to do anyway.

Doing ethics

We can apply Latour's and Flyvbjerg's ideas to further develop the view on ethics as *doing ethics*: as facilitating processes of ethical reflection, inquiry, and deliberation. We can recognize this as a constructivist stance, which holds that people construct morality through social interactions and processes. This is different from a realist stance, in which people study phenomena as if they exist externally. It is also different from engineering understood very narrowly, as creating a solution for a problem that is given and fixed; a problem that cannot be questioned.

Rather, we can understand *doing ethics* as facilitating two processes that, ideally, need to happen iteratively: a process of *problem-setting*, of understanding, framing, and articulating a particular problem in its context; and a process of *solution-finding*, of envisioning, building, and evaluating potential solutions. *Doing ethics* can complement the natural and social sciences. The sciences can focus on finding facts, while ethics entails a process of reflection, inquiry, and deliberation. People from different disciplines can collaborate and use these facts to better understand the problem at hand, preferably from different angles, and to envision and create potential solutions, using knowledge from different domains. Moreover, *doing ethics* can complement technology. People with technology backgrounds can collaborate with people from other backgrounds and jointly engage in a process of ethical reflection, inquiry, and deliberation, in order to better align their work with values and concerns in society.

This would be rather different from what sometimes happens in technology development and innovation, at least in my experience. People from different disciplines find it hard to work together, or if they work in one project, they work on isolated tasks. As a consequence, they cannot combine their respective fields of expertise. Typically, discussions focus on *facts*, so that there is little room to discuss *values*; for instance, the concerns or needs of the people who may be affected by the innovation. Or the people involved tend to focus on *means*, on developing and deploying technology, so that there is little room to discuss *ends*; for instance, a practical desired state of affairs that the project aims to contribute to realizing.

Let me mention a positive exception. This was in an international research project (TA2), in which we created prototypes for telecommunication and multimedia applications that would promote togetherness between family members or friends. At the start of each meeting, the project manager presented the same sheet, with the project's mission and assumptions, and made room to question and discuss these. This helped to create a shared understanding of the project as a whole. In addition, he actively promoted collaboration between people working on technology and people working on user experience, and an iterative approach, in which we created a series of demonstrators and prototypes and evaluated these together with prospective users, in a series of focus groups, experiments, and pilot studies.

Also, in my experience, it can be hard to make iterations. Of course, many are familiar with agile processes, in which you go back and forth between design and evaluation, typically in close collaboration with clients or prospective users. However, it happens less often that there is room to question or discuss the project's basic assumptions: its definition, scope, and goal. Project managers and clients typically find it easier to stick to plans they negotiated and agreed upon and find it harder to make room for profound questioning of the starting point and finish line they established.

Let me mention another positive exception. In another international research project (WeCare), we developed several services to support people who provide informal care to people who suffer from dementia, often their parents. In one meeting, the project manager questioned the 'return on investment' of our project. The money we spent on research and innovation versus the benefits for the prospective users of these applications. Maybe the money could be better spent on improving or upscaling already existing support programs for these people? Or maybe we could enlarge our scope and look at other problems, maybe in other countries, where the money could be spent on more distressing situations, with a larger return on investment, with more 'bang for the buck'.

Although I do endorse, in general, questioning and iterations, I do also appreciate the need to channel such questioning and to manage such iterations. I would, for example, not endorse random questioning or continuous iterations that go back to square one and start from scratch. Rather, I would propose making room for iterations in an organized manner, see Chapters 17–19.

Greek letters

In Dutch, my mother tongue, the different domains of knowledge are referred to with Greek letters: *alpha* for the humanities, *beta* for the natural sciences, and *gamma* for the social sciences. With a bias towards European culture, we can say that the humanities came first, with the ancient Greeks, for whom all knowledge was (natural) philosophy. With the European Enlightenment came the natural sciences, for example, *The Royal Society*, the first organization to promote modern science, was founded in 1660. And around 1900, the social sciences emerged; the first lab for psychology experiments, for example, was founded by Wilhelm Wundt in 1879, and *The Interpretation of Dreams*, the book that made Sigmund Freud famous, was published in 1899.

We can continue the Greek alphabet with *delta*, which we can use to refer to the field of design and technology. Interestingly, design and technology deploy a particular mode of reasoning.

The main modes of reasoning in science are *induction*, which goes from specific observations to a general pattern (*copper expands when heated + iron expands when heated → metals expand when heated*; in computer science, machine learning uses algorithms of this kind), and *deduction*, which goes from premises to a conclusion (*all persons are mortal + Socrates was a person → Socrates was mortal*; in computer science, symbolic reasoning uses algorithms of this kind).

Now, people in design and technology can use these modes of reasoning. In addition, they can use *abduction*: a mode of reasoning that flows in the opposite direction: from ends to means; from conclusion to premises, if you like. They start with envisioning a desirable end state, for example, a recreational tent that is both spacious and lightweight, and then envision and create means to achieve that end state, for example, a light, waterproof fabric and flexible tent poles with carbon fibre, to make an igloo shape (*light, waterproof fabric + flexible poles, in igloo shape ← spacious, lightweight tent*). In practice, they will often do this in an iterative manner, moving back and forth between design and evaluation. Interestingly, the symbol for *delta* (Δ) refers to difference; in this case, the difference between what-is and what-could-be. The design and application of technology involve understanding what-is and creating what-could-be.

A three-step process

Let us turn to a practical method. Over the years, I have developed the following three-step approach to facilitate ethical reflection, inquiry, and deliberation, in a collaborative and iterative process:

Identify issues that are at play in your project, currently or potentially, for example, outcomes of the project that can be problematic and reflect on these. A handful of issues works best; if you have more, you can cluster them; if you have less, you can explore more. Later on, we will discuss virtues like justice and courage, which can help to identify issues and to speak up.

Organize dialogues with the people involved and with relevant stakeholders, both within your organization and outside your organization, to inquire into these issues from diverse perspectives and to hear diverse voices. Later on, we will explore how virtues like curiosity and creativity can help to create an open atmosphere for such dialogues, in which people can collaborate.

Make decisions, and try-out, and evaluate these decisions in small-scale experiments, and be transparent and accountable about the findings. The key is to steer your project consciously; to act consciously and carefully. Later, we will discuss how virtues like honesty, self-control, care, and collaboration are critical for organizing such experiments.

In later chapters, we will explore ways to organize iterative and collaborative processes (in Chapter 17 on Human-Centred Design), to learn about diverse actors' and stakeholders' values and societal concerns (in Chapter 18 on Value Sensitive Design), and to promote anticipation, responsiveness, diversity, and reflexivity (in Chapter 19 on Responsible Innovation).

I have found that this three-step process can help people to integrate ethics in their project through a series of relatively small steps. It is more practical than an extensive ethical analysis or technology assessment. Also, it is meant as an alternative to a view in which ethics is perceived as a roadblock, as a barrier to innovation: *Stop; you cannot go further.* I would like to replace that image of a roadblock with a steering wheel in a vehicle. That image highlights project team members' agency to steer their project. They can use the wheel to steer their project in a direction that is aligned with needs and concerns of society, and to safely reach a desirable destination. The steering wheel can prevent taking a wrong exit or hitting the guardrail.

You can follow this three-step process in various ways, depending on your expertise and skills, the available budget, and lead times. You can do a workshop with three or four people (or more, when working in sub-groups), in only 60–90 minutes. This will already yield useful results. It would be silly to expect sophisticated answers to very complex questions in a session of 60 minutes. Rather, such a workshop can yield insights and can help to articulate topics or issues more clearly and precisely, so that you can address them in another meeting, for example, with the project's client.

Crucially, there is a step zero that needs to precede these three steps; to describe the application, product, service, or system that the project is working on, in very practical terms.

We need to understand the innovation, its functioning, and outcomes practically, in order to identify and discuss issues. Prior to a workshop, one or more project team members can create a textual or visual description of the innovation, from the perspective of an operator or user: *What does it do, practically? How is it different from currently available systems? And how does the application affect other people or groups?* I have found that asking simple questions like these can help project team members to describe what they work on and hope to achieve in practical terms. Without such a description, they can tend to talk about their innovation in rather nebulous terms.

Let us take this three-step approach for a spin. Think of a project that you work on, worked on, or will work on. Can you think of ethical or societal or other issues at play in this project? Maybe you have second thoughts about something? Maybe someone mentioned a sensitive issue? Can you imagine organizing dialogues about such issues? How would you do that? Whom would you invite? What would you need to pay attention to? How would it feel to find ways to proceed with these issues? To make progress in your project, and also act consciously and carefully?

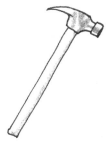

Chapter 3

Is technology a neutral tool?

We explored the notion of *doing ethics*, of promoting ethical reflection, inquiry, and deliberation in a collaborative and iterative process. Now, before we can apply *doing ethics* to the design and application of technology, we need to look into several questions concerning technology. We will explore whether technology is a neutral tool. I will argue that technologies are like tools, that we can use them for different ends and in different ways, but that they are *not* neutral. People and technologies form complex webs of interdependencies, which leads to other questions: *How do we interact with technology? And how do we want to interact with technologies and with others?*

DOI: 10.1201/9781003088776-4

Let us start with exploring whether technology is neutral. Take, for example, a car. You may think that a car is a neutral tool that one uses to travel from A to B. If, however, we zoom-out and look at the history of how cars were developed and adopted, we see many factors that affected their development and adoption and many issues that were, in turn, affected by the development and adoption of cars. Around 1900, electric battery-powered cars and combustion engine-powered cars both existed and a number of contingent factors affected their respective development and adoption. The discovery of deposits of 'cheap' fossil fuel promoted the development and adoption of fossil fuel powered cars. They won the race, so to say, not the electric cars. Please note that 'cheap' refers to low, short-term financial costs, not to the huge, long-term costs of harms to ecological and social 'externalities' that the extraction and burning of fossil fuel entail. Similar problems are found, for example, in training large language models and mining crypto-assets. They use lots of energy, often from fossil fuel, and thereby contribute to global warming, whereas these applications' benefits are probably not proportional to these environmental and social costs.

In turn, the broad adoption of gasoline cars made the fossil fuel industry increasingly important and had huge impacts on geopolitics in the 20th century, motivated international conflicts, and has grown to become a major cause of the current climate crisis. The *car culture* (or *car cult*) has also had a huge impact on urban planning in, for example, the US, or rather, the lack of urban planning. Many places in the US suffer from *suburban sprawl*: untethered growth of buildings over large areas, with diverse adverse social and ecological outcomes. People are effectively forced to use cars to go anywhere; many daily destinations, like shops or schools, are beyond walking distances. I can contrast this with the Netherlands, which has dedicated bike paths, which enable people to use bikes for short trips, and an excellent public transport system, with rental bikes at most train stations.

We make all sorts of decisions, which lead to the design and application of certain technologies, rather than alternative technologies. And the adoption of those technologies that 'win', in turn, make specific, further decisions more likely than other decisions. This is known as *path dependency*. Once there is a system of gasoline stations in place, using gasoline cars is easier than using electric cars. If we would want to promote the usage of electric cars, governments, companies, and citizens will need to make efforts. Governments can offer tax cuts and subsidies. Companies need to take risks and manufacture electric cars, or invest in building electrical charging stations. Citizens will need to say goodbye to their gasoline cars, use electric cars, and form new habits, for example, for charging.

People and technology

'We shape our tools and thereafter they shape us', said John Culkin in an interview with media scholar and critic Marshall McLuhan. Sometimes this shaping happens unintentionally. It can then be worthwhile to make this process of shaping more explicit. Sometimes the shaping happens intentionally. Some technologies are designed to bring about change in society, to shape society. Maybe you agree with these objectives. But if you do not, then you may want to question this shaping process.

A classic example of the application of technology to have effects on society comes from Langdon Winner, a pioneer of *Science and Technology Studies*, a field in which people study how scientific facts and technological artefacts are created and used, with special attention for the role of political and economic power. Winner discussed the work of Robert Moses, who held key positions in urban planning in New York City from the 1920s to the 1960s. Moses ordered the building of a series of bridges across parkways. These bridges were so low that they prevented buses from going underneath. Effectively, these bridges prevented poor people, who typically could not afford cars and depended on buses, mostly people of colour, from visiting the parks and beaches. Moses used technology to achieve racist outcomes.

Another, less consequential example is of shaving devices. Typically, shavers for men used to have screws, which suggests that men are able to do maintenance or repairs. Shavers for women, in contrast, have no screws, which suggests that they are unable to maintain or repair these devices. Notably, most shavers now on the market have no screws, which suggests another story: of buying a new shaver instead of trying to maintain or repair it. These features are referred to as *scripts* or *affordances*; these prescribe, to a smaller or larger extent, what people can and cannot do with these products. You can also think of the default settings in a mobile phone app; they are meant to steer people's behaviour. Although, with some effort, you may be able to override these settings and use the app differently from the *script* that the developers put into it. A notification beep on your smart phone nudges you to check your phone. You can ignore it, but that will require effort.

Cars, bridges, and shavers, and indeed all technologies, are *not* neutral. An instrumental view on technology is mistaken; we cannot use technologies as neutral instruments. A deterministic view is also erroneous; technologies do not determine our behaviours. Rather, the design and usage of technologies are embedded in a web of more or less intentional decisions, of both designers and users, and these decisions have effects on the ways in which people use these technologies.

Historian of technology Melvin Kranzberg described the relationship between people and technology succinctly: 'Technology is neither good nor bad; nor is it neutral'.

Interacting with technologies

You can relate to technology more or less consciously. Less conscious relating would entail taking things for granted as they appear to you, for example, checking your mobile phone every time it buzzes. More conscious relating would entail, for example, questioning current habits and creating alternative habits. You could, for example, zoom-out and see the connections between your usage of your mobile phone and the companies collecting your personal data and extracting value by selling ads to target you. This may motivate you to use your mobile phone differently.

We can dive deeper into the relationship and interactions between people and technology. Dutch professor of philosophy of technology, Peter-Paul Verbeek, drawing from the work of Don Ihde, distinguished between different types of 'mediation' between people and technology, which we can understand as different ways of interacting with technologies:

Embodiment or cyborg: The technology has become transparent, like glasses that help you to see better and which you become unaware of over time; in a cyborg relationship, the technology merges even further with the human body, as a prosthesis or brain implant does;

Hermeneutic or augmentation: The technology represents an aspect of the world, like a thermometer that shows you the temperature; or, for example, an Augmented Reality app on a mobile phone that provides information by adding it to your view on the world;

Background or immersion: You experience the effects of a technology, but cannot easily perceive its operation, like a heating system, where you feel heat but cannot see how it works; many 'smart home' systems are designed to blend into the background in this way;

Alterity or agent: You interact with a system as if it were an 'other', for example, a ticket machine with a touch screen, which asks questions, which you answer; such a system can be further developed into an agent that performs tasks more or less autonomously.

Now, if we look, for example, at so-called 'smart devices', we can see how different types of interaction are possible, either by design or through usage. Often, we will experience these types of devices as being part of the *background*, for example, when they sit on a table in the corner of the living room. Designers can create an experience of *immersion*, so that you take them for granted, as if they were a natural phenomenon. Now, if the device relays information about the world, for example, a weather report, the interaction becomes *hermeneutic*. A pair of 'smart' glasses may project the advice 'bring an umbrella' when you approach your front door: a form of *augmentation*. When you wear such a device on your body, for example, as a wrist band, the relationship becomes one of *embodiment*; or maybe even *cyborg*, when you wear it as if it has become a part of your body. And the relationship to such a device can become one of *alterity*. The first thing that many people do, waking up, is look at their smart phone, and the last thing they do before sleep, again, is look at their smart phone. Over time, and with each interaction, the smart phone can become something like an *agent* to which you relate.

These examples draw attention to the role of both designers and users in shaping the relationships and modes of interaction between people and technology. These relationships and interactions are not accidental. They are also not inevitable. Very often, these interaction patterns are designed intentionally, in order to promote specific behaviours to achieve specific goals.

Critique

Many social media apps, for example, are designed to maximize screen time, using decades of knowledge from the domain of gambling and slot machines. With buzzing sounds and flashing icons, they grab our attention as often as possible. With infinite scrolling and seductive suggestions ('watch next'), they hold our attention as long as possible. 'The best minds of my generation are thinking about how to make people click ads', lamented Jeff Hammerbacher, former data lead at Facebook.

Former Google designer Tristan Harris is one of the people who helped to expose such practices. We are vulnerable to how these designs activate our behavioural impulses. Our brains have evolved to be activated by novel or interesting sensory input. This was helpful when we used to walk in the proverbial savanna and needed to be aware of dangerous predators and of scarce food stuffs. But very unhelpful in our current media environment. Similarly, it used to be wise to grab anything sugary and fatty, millennia ago; it is, however, unwise in a supermarket.

Many so-called 'social media' platforms function *not* as tools for people to connect to other people or to content that benefits you. Instead, they are tools for companies to grab and hold people's attention and sell advertising opportunities to whoever wants to influence people's behaviour. People like Jaron Lanier (in *Ten arguments for deleting your social media accounts right now*), Douglas Rushkoff (in *Team human*), and Sherry Turkle (in *Reclaiming conversation*) argue that social media platforms are designed to corrode meaningful conversations and instead promote so-called 'engagement': interactions that are based on outrage-fuelled and outrage-fuelling algorithms, which results in echo chambers, filter bubbles, and polarization.

Another line of critique is aimed at the ways in which algorithms can propagate or exacerbate existing injustices. This is often referred to as 'bias', which needs to be mitigated or prevented in data sets and in machine learning. This critique is brought forward, for example, by danah boyd, founder and president of *Data & Society Research Institute* and visiting professor at New York University; Meredith Broussard, data journalism professor at New York University; Safiya Noble, associate professor at UCLA and co-founder and co-director of the UCLA *Center for Critical Internet Inquiry*; and Timnit Gebru, co-founder of *Black in AI* and founder of the *Distributed Artificial Intelligence Research Institute*. They critique privileging the experiences and concerns of, typically, white men, at the expense of the experiences and concerns of black people or of women, or, more generally, of people with diverse skin colours and diverse gender roles.

Their efforts are sometimes referred to as *Critical Data Studies*, a field that questions and criticizes assumptions that are commonly held but incorrect, notably about data as representing facts, and algorithms as being objective. Anytime you hear someone talk about data as 'facts' or about algorithms as 'objective', you can ask questions like: *Where did the data come from? Which types of data were used? Which were not used? Which tools were used to collect and interpret the data? Which types of data were therefore* left out *and which interpretations were* not *used? Which tools were used to create the model and the algorithm? Which biases therefore became part of the model and of the algorithm?* We will discuss this topic further in Chapter 11 on duties and rights.

Have you noticed how people, including yourself, talk about technology? How they think or feel about technology? How they relate to it, interact with it? As an exercise, you may want to take one project and try to find out how people in that project think or feel or talk about the technology they work on. Do they view it as something good? As something bad? As a neutral tool? You may want to listen carefully to learn about their assumptions and commitments. And what about you? How do you think and feel about this technology? How do you envision people interacting with the technology that you are helping to design or apply?

Chapter 4

Value, wellbeing, and economics

Where have we got so far? We discussed a process view on ethics and outlined a three-step process for doing ethics. In addition, we discussed the notion that technologies are not neutral. Currently, we will explore the notion of *value*. Technologies can be used to generate value, extract value, or to distribute value. We will zoom-out and explore how the design and application of technologies can be used to promote wellbeing and/or economic development.

Let us start with a key question: *What is technology for?* Philosophers have debated that question and came up with complex answers. Others will give relatively simple answers. A saw is for sawing. A hammer is for hammering. Okay. But you

DOI: 10.1201/9781003088776-5

work on more complex technologies, maybe a machine learning system, a digital product or device, or an online service. *What are these for?* What is their value? I guess that depends on whom is asked that question. Some would answer: 'To make processes more efficient, to sell things, to make profits'. Others, like myself, would reply: 'To create conditions in which people can live well together, and flourish'.

The last couple of years we have seen a 'tech lash'. A growing group of people worry about the negative consequences of technologies on society, on daily lives, and on wellbeing. This type of critique is not new. There have been many critics of technology in the 20th century. Lewis Mumford (1895–1990) critiqued mass production and consumerism. Günther Anders (1902–1992) critiqued mass media and drew attention to the nuclear threat. Jacques Ellul (1912–1994) pushed back against the focus on efficiency and rationality. Ivan Illich (1926–2002) criticized institutional approaches in, for example, education and health care and envisioned 'tools for conviviality', tools that can help us to live well together. Hannah Arendt (1906–1975) discussed the corrosive effects of technology on the human condition, notably on our abilities to act as citizens. Hans Jonas (1903–1993) pioneered the *imperative of responsibility*, which was adopted by environmentalists. More recently, Hubert Dreyfus (1929–2017) critiqued overly positive views on AI (*What Computers Can't Do: The Limits of Artificial Intelligence*). Donna Haraway (born 1944) developed a feminist view on technology and questioned dominant assumptions about boundaries between people and machines (*A Cyborg Manifesto*). Neil Postman (1931–2003) critiqued media (*Amusing Ourselves to Death*) and technology (*Technopoly: The Surrender of Culture to Technology*). Langdon Winner (born 1944) discussed the ways in which technologies are used to exercise power (*Do Artifacts Have Politics?*).

Currently, a growing group of scholars and scientists, journalists and activists, entrepreneurs, and whistle blowers draw attention to the downsides and dangers of technology.

Relatedly, they critique untethered neoliberalism and corporate greed, unsustainable modes of production and consumption, (neo)colonialist and racist practices, typically based on extraction and exploitation of nature and of 'others', and systems that propagate or exacerbate inequalities and polarization.

With systems thinking, we can see that technology often plays dual roles: both as a cause and as an effect. Untethered neoliberalism and corporate greed, for example, are enabled by technologies for collecting data and selling ads and also enable the further deployment and intrusion of such technologies, resulting in vicious reinforcing feedback loops that are hard to stop.

Now, let us explore two ways in which technology can be used to generate or extract or distribute value, in the next two sections: in terms of wellbeing and in terms of economics.

Wellbeing

One way to understand the value of technology is in how it can be used to promote people's wellbeing. Discussions about wellbeing often make me think of this parable with the elephant and the people with blindfolds. There is an elephant. Five blindfolded people stand around it. Each person touches one part of the animal and expresses what they believe it to be. One feels its trunk and believes it is a large and muscular snake. Another person feels its ear and declares that it is a thick and hairy plant leaf. Someone else feels the elephant's leg and believes it is a tree.

Similarly, different people can have disparate answers to the question: *What is wellbeing?*

A widespread belief regarding wellbeing is that it consists of owning or consuming material goods. Of course, there is a minimum of material goods needed for wellbeing. Every person needs housing, food, health care, education, employment, family life, and social life in order to flourish. Furthermore, material goods can enable people to engage in meaningful and fulfilling activities, especially if they involve: using and developing one's talents and skills, contributing to a greater goal, creating and nurturing relationships with people one cares about, or improving one's health. Beyond some level of material welfare, however, these material goods begin to matter less. And there is a level beyond which too much focus on material goods can impede one's wellbeing. Moreover, a focus on accumulating wealth can easily lead to economic inequalities, which can lead to tensions in society, so that all people's wellbeing is undermined.

We can trace many views on wellbeing in Western culture back to ancient Greece. Aristotle (384–322 BCE) proposed that wellbeing (*eudaimonia*) entails the development of people's potential. He understood wellbeing as *flourishing*, as living a meaningful and fulfilling life, as a social, life-long, and creative activity. More on that in Chapter 15 on virtues and flourishing.

Interestingly, Aristotle's view on wellbeing has informed positive psychology, a field of psychology that emerged around 2000, and which aims to understand and create conditions in which people can engage in positive activities and have positive experiences. A pioneer of this field, Martin Seligman, understands wellbeing as consisting of the following five elements: positive emotions, like joy; engagement, for example, in a creative activity; relationships, notably, with family or friends; meaning, for example, by contributing to societal goods; and accomplishment, like achieving results in challenging activities.

Another view on wellbeing with roots in ancient Greece was articulated by Epicurus (341–270 BCE). He advocated pursuing happiness by engaging in activities that provide pleasure and by avoiding activities that cause pain. His approach is called *hedonic*, but it is rather different from the current, colloquial meaning of hedonism. Epicurus advocated enjoying relatively simple daily activities, like eating, being with friends, and the pursuit of economic independence and resilience. He famously wrote in a letter to a friend: *Will you please bring cheese, so that we can have a feast?* We will discuss modern interpretations of this view in Chapter 9 on consequences and outcomes.

Economics

Another way to understand the value of technology is in terms of economics. And here we need to talk about the elephant in the room: the dominant paradigm of growth, the belief that growth is always beneficial and desirable, needs to go unchecked and needs to be promoted at all costs.

National governments often use Gross Domestic Product (GDP) as a measure for how well they are doing. GDP is the monetary sum total of all the goods and services produced by a country in a given year. Over the years, the idea has taken hold that a high and growing GDP is good. This is, however, fundamentally unwholesome, since any monetary spending, regardless of its nature, contributes to GDP. An environmental hazard, like an oil spill, contributes to GDP because money is spent on cleaning up the mess. When people buy unhealthy food and get sick and pay for their medical treatment, GDP increases two times: for buying the unhealthy food and for paying the medical treatment. In short, GDP is not the best indicator for wellbeing.

There used to be a time when GDP made more sense. It was introduced in the 1930s, during the depression, when many people lacked basic goods, when economic growth made sense. It was also useful after the Second World War, for rebuilding countries and economies. Currently, however, in a world that needs sustainable development and a just distribution of wealth, a narrow focus on GDP makes very little sense. Some may argue that economic growth is still a good idea because wealth would eventually 'trickle down', from rich people to poor people. This view, however, has been proven wrong. Sadly, economic growth often exacerbates economic inequality. Often, economic growth enables powerful and rich elites to get even more powerful and rich and to further exploit and harm people with less power and wealth.

We can, by the way, draw a parallel between the usage of GDP and the ways in which data scientists use indicators in order to optimize some algorithm. The question is whether that indicator is (not) an adequate pointer for what is important, for what needs to be optimized.

Regarding GDP, a growing group of economists, politicians, and NGOs have advocated using alternative metrics; commonly referred to as 'beyond GDP'. In 1990, the United Nations launched their yearly *Human Development Reports,* based on the work of Amartya Sen (see next section), with, for example, an *Inequality-adjusted Human Development Index,* which includes a range of factors that need to be in place in order to enable people to flourish, for example, employment, health care, and education. Similarly, and more recently, Scotland, New Zealand, Iceland, Wales, and Finland launched the *Wellbeing Economy Alliance (WEAll),* to promote a focus on human wellbeing, understood broadly. Notable is also the work of the *Happy Planet Index,* introduced by the *New Economics Foundation.* Simply put, this index is calculated as people's life satisfaction ('happy') divided by ecological footprint ('planet'). This index combines concerns for wellbeing and for sustainability.

Capability approach

There is an approach that I have found particularly useful to understand the ways in which technology can promote both people's wellbeing and sustainable development: the Capability Approach (CA). It was developed by Amartya Sen (born 1933), economist and philosopher, and recipient of the 1998 Nobel Memorial Prize in Economic Sciences, and Martha Nussbaum (born 1974), professor of law and ethics at the University of Chicago.

The CA aims to understand and create conditions in which people can flourish. More specifically, it aims to enable people to develop and exercise relevant capabilities, so that they can 'lead the kind of lives they have reason to value', as Sen put it. Key capabilities include, for example, the capability to live with good health, to be adequately nourished, to have adequate shelter, to move freely from place to place, to be secure against violent assault, to live with others, to engage in social interaction, and to participate effectively in political choices that govern one's life. The CA puts human dignity and human autonomy centre stage.

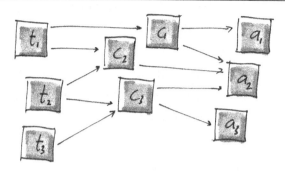

The basic structure of the CA can be visualized as three groups of boxes that are connected by bundles of arrows, from left to right. On the left are technologies, which we understand as resources or as tools. In the centre are human capabilities, the focus of this approach. On the right are people's practical ways of living and activities. Ideally, people are enabled to access and use technologies in order to expand and exercise relevant capabilities and in order to pursue activities that contribute to their flourishing, so that they can live well, together with others.

Understanding the arrows between these boxes can help to steer clear of two pitfalls that can easily arise when we think about using technologies to promote people's flourishing.

First, there is the pitfall of focusing too much on the technologies and forgetting the broader context in which these technologies are used. The CA draws attention to a bundle of arrows that go from technologies to capabilities: these arrows represent various *conversion factors*, which need to be in place in order for a technology ('merely a means') to enable people to actually expand and exercise relevant capabilities. Conversion factors can be personal, social, or environmental.

A project involving the development of a podcasting service for information on health and cattle management in a rural area of Zimbabwe can help to illustrate these conversion factors. This example comes from the work of Ilse Oosterlaken, who pioneered the application of the CA in the domain of technology, innovation, and design. In this project, a voice-based technology was chosen to address the *personal* conversion factor of many inhabitants' illiteracy and reliance on spoken texts. The system used speakers rather than headphones in order to align with the prevalent social practice of 'sitting and sharing under a village tree', a *social* conversion factor. And the system used solar panels to recharge its batteries to tackle the *environmental* conversion factor of an inadequate electrical grid. The CA can help to understand people's contexts and conversion factors that need to be included in the design process, in order to enable people to extend and exercise relevant human capabilities. Rather than narrowly focusing on technology.

The second pitfall is the endorsement of specific behaviours. The ambitions and assumptions of the people working on the project can lead to them creating products or services that are biased towards their own ambitions and assumptions. In a way, they will endorse specific behaviours and, thereby, hamper people's freedom of choice, and thus their wellbeing. Crucially, the CA understands capabilities as freedoms. In the podcasting example, designers can promote freedom by enabling people to use the system also for other types of information or by providing internet access in a community facility. People can then use the internet for diverse purposes, like commerce or learning. The CA can help to ensure that people can indeed freely choose to live different versions of the good life, rather than restrict their freedom.

The CA can help you to focus your project on promoting people's flourishing, rather than focus on technology, to understand and take into account conversion factors, and to enable people to expand relevant capabilities. Significantly, this includes the capability to form conceptions of living well, of the good life, to engage in critical reflection, and to actively plan their lives.

The CA will reappear in Chapter 18 on Value Sensitive Design, where it will be the basis for a specific approach to planning and organizing innovation projects.

Sustainable development

I would like to briefly discuss the ideas of Kate Raworth and Mariana Mazzucato; they propose to think about economics, development, and innovation in more sustainable ways.

Kate Raworth is the author of *Doughnut economics: Seven ways to think like a 21st-century economist* (translated into over 20 languages). She invites people to move away from the unsustainable premise of infinite growth on a finite planet and to move towards sustainable strategies and policies. Our narrow focus on growth is like an addiction, which we need to replace with healthier habits. She visualizes her argument with two concentric circles: the outlines of a doughnut. We need to be in the safe space between these circles; they function as ecological and social boundaries. We need to *not* overshoot the larger circle, the ecological boundaries of our Earth's capacity to support life, notably in terms of a stable climate, biodiversity, fertile soils, and healthy water. And we need to *not* undercut the smaller circle, the social boundaries with regards to human dignity, in terms of, for example, having access to food, housing, health care, education, and political rights. In *Doughnut economics*, Raworth discusses a history of ideas in economics. She reminds us that Simon Kuznets, who helped to establish the GDP, warned about its limitations: 'Distinctions must be kept in mind, between quantity and quality of growth, between costs and returns, and between the short and long term ... Objectives should be explicit: goals for "more" growth should specify more growth of what and for what'.

Raworth also discusses the harms of thinking in terms of *externalities*. Businesses routinely choose to draw a dotted line and to not bother about what happens outside that dotted line: where and how a company obtains raw materials, for example, in toxic rare earth metals mines in conflict areas; which people do labour, often in low-income countries, in unhealthy sweatshops, with little regulation for working conditions; and about waste, pollution, and emissions, materials in landfills, microplastics in oceans, and greenhouse gases in the atmosphere. If you treat all these factors as externalities, you are bound to transgress the ecological and social boundaries we need to stay within.

Drawing from systems thinking and the tradition of sustainable development, Raworth proposes to make economics *regenerative*, rather than extractive, and *distributive*, rather than unjust.

Mariana Mazzucato pioneered the idea of *mission-oriented innovation*, in a series of popular books and influential reports. In *The entrepreneurial state* (2013), she debunked the myths that the private sector is where innovation happens, and that the public sector is best kept away from innovation. The opposite is true. The public sector has been critical in many innovations that have shaped our lives. The US government, for example, invested in key technologies and helped to start entire industries, like computers (with NASA's moon missions), the internet (DARPA), and GPS (Department of Defence). The government helped to found *Tesla Inc.*, with a $465 million loan from the Department of Energy, and continues to support it, through various financial incentives and tax cuts. In *The value of everything* (2018), Mazzucato investigated the structures and processes of value creation and value extraction ('making and taking in the global economy' is the book's subtitle). Too often and harmfully, the *extraction* of value, for example, through owning property that one can extract rent from, as is the case for most activities in the financial sector, is rewarded much more than the *creation* of value, which happens, for example, in producing goods and offering services in the real economy. Significantly, many industries are increasingly *financialized*; they are managed with the logic of the financial sector: to maximize short-term financial profits, rather than facilitate a sustainable business model that works on the long term. Private equity funds play a key role here. They buy companies, cut away so-called 'unnecessary' parts, make them profitable on the short term, on paper, and then sell them with a profit. Very often, however, these 'unnecessary' parts are of critical value, so that these companies' performances dwindle and their employees, customers, and suppliers are worse off. In *Mission economy* (2021), Mazzucato builds an enticing vision for organizing innovation, which she has helped to implement in various places, notably in the European Union's 95.5-billion-euro *Horizon Europe* funding program for research and innovation. The program is organized to address and help solve 'grand societal challenges', like making cities climate neutral, promoting healthy soil and food security, and making oceans, seas, coastal, and inland waters healthy. Mazzucato's view is based on a realistic and positive view on the interplay between government and industry. She proposes seven principles for a new political economy to steer innovation: value is created collectively, *not* solely by industry; markets need to be shaped proactively, *not* fixed reactively; innovation requires dynamic capabilities, like learning, in the organizations involved; finance can help to steer innovation, notably, in the form of outcomes-based budgeting; risks and rewards of innovation need to be distributed fairly, *not* with society bearing the costs and companies reaping the profits; with partnerships organized towards purpose and value creation; and with open systems in which citizens can participate actively in the innovation process.

Hope

We explored some ways in which innovation and technology can help to create value, in terms of wellbeing and economics. We discussed the need to go 'beyond GDP' and towards sustainable business models. These explorations are not frivolous. We face multiple complex and global challenges. There are moments when I find these challenges very daunting.

Maybe another elephant can help. The elephant that represents wisdom and perseverance. Elephants are intelligent. They are powerful, with their sheer size and weight. Yet they are sensitive, with their flexible and agile trunk. When you want to do good in your projects, you will need to cultivate power and sensitivity. And perseverance, because change requires time.

Maybe the image of an elephant can also help to cultivate hope: to persevere in adverse conditions. Significantly, Raworth closes *Doughnut economics* with a reference to hope: 'an optimistic vision of humanity's common future', and in the last chapter of *Mission economy*, Mazzucato expresses the 'hope [that] this book has given us some of the tools to imagine' another, more sustainable and equitable world.

Now, after our discussions of ethics, of technology, and of wellbeing and economics, it is about time to get to the core: different perspectives on doing ethics. Before that, however, we need to briefly discuss three topics that regularly feature in discussions of technology and ethics: the *trolley problem*, which we will engage with creatively; *privacy*, which we will discuss in a broad sense; and *responsibility*, which we will understand in terms of knowledge and agency.

Chapter 5

The problem with the 'Trolley Problem'

In 1967, English philosopher Philippa Foot (1920–2010) wanted to discuss the 'doctrine of double effect': the difficulty of having to evaluate acting out of good intentions vis-à-vis bringing about harms as side effects of this acting. She invited readers to imagine being in the position of 'the driver of a runaway tram which he can only steer from one narrow track on to another; five men are working on one track and one man on the other; anyone on the track he enters is bound to be killed'. She wanted to illustrate the difficulty of such dilemmas. It is quite unlikely that she wanted to present a problem that computer scientists or software engineers can solve through calculation and optimization.

DOI: 10.1201/9781003088776-6

In the years since, however, people have come up with a range of variations of what has become known as the *Trolley Problem*, where you are invited to imagine being in the position of a person standing next to the track, with a lever at hand, which you can use to divert the trolley: to kill one person instead of five. In the *Fat Person* variation, you stand on a foot bridge, together with a large person; you can choose to throw that person from the bridge to stop the trolley underneath. Many people shy away from throwing this person; it feels like directly killing that person, rather than pulling the lever and having the trolley do the killing.

The thought experiment has inspired scores of psychology experiments, notably MIT's *Moral Machine*. This was a large-scale, online survey in which people were invited to decide what a self-driving car ought to do in case of an impending accident. Millions of people participated in this survey. They were given a series of pairs of scenarios, presented as map-like diagrams, with various numbers and types of pedestrians and of passengers. For each pair of scenarios, they had to choose between, for example, driving ahead and killing pedestrians, or veering off into a roadblock and killing passengers. Based on 40 million responses from people from over 200 countries, they found several general preferences, like sparing humans over animals, sparing more lives over fewer lives, and sparing younger people over older people. They also found differences between cultures. People from countries with collectivistic cultures prefer sparing lives of older people over younger people. And people from countries that are poorer and have weaker institutions tend to spare pedestrians who cross illegally.

What can we learn from such experiments? Can they give us clues for programming self-driving cars? Maybe. Or maybe not. We need to admit that this was a *psychological* survey. It probed people's preferences. Scottish philosopher David Hume (1711–1776) famously reminded us to not confuse *is* and *ought*: you cannot derive moral statements from empirical findings. If people do X or say that they would do X, it does not automatically follow that X is morally acceptable or recommendable. This is called the *naturalistic fallacy*: the false belief that what *is the case* also *ought to be*.

Also, the *Moral Machine* survey used scenarios that depict different types of pedestrians: homeless people, with a baggy coat; business people, with a suitcase; medical personnel, with a first aid kit; and criminal people, with a mask and loot. Participants probably took these differences into account in the survey. It would be undesirable to require of people that they carry such identifiers in public spaces, so that self-driving cars can take their identities into account. Or will pedestrians be required to wear tags, so that self-driving cars can identify them?

The problem with such takes on the *Trolley Problem*, I believe, is that people can think that these can help to 'solve' ethics: to reduce ethics to calculation and optimization. Instead, I will present an alternative take on the *Trolley Problem*, possibly more aligned with its original function as a thought experiment: to promote ethical reflection, inquiry, and deliberation.

First, we will widen our view. There is more at play than the trolley, the workers, and you. Then we reflect on our agency as professionals. You can do more than pull a lever. The original description is parsimonious on purpose: a trolley with a malfunctioning brake, a track with five people, another track with one person, and you with a lever. Real life is more complex.

Why is there no klaxon on the trolley, to warn people on the track? And why did the brakes malfunction? Was there a flaw in the design of the brake system? Did the company examine past accidents, to learn from them? Or was it lack of maintenance? And how did that come about? Was the maintenance budget insufficient? Or were maintenance experts made redundant? Speaking of priorities: How does the trolley company spend their budget? Do they pay fair salaries to their employees? Are they motivated or rewarded to report errors or accidents? Or do they work with subcontractors who hire 'self-employed' professionals or 'gig workers'? How is their work morale and capacity to improvise? Did the driver get sufficient rest between shifts? How are their working conditions? What health and safety policies are there for the people who work on the tracks? Or, zooming-out, we can ask: Why did the state privatize the trolley services? How is competition regulated between different service providers? Are there incentives for them to collaborate on safety? Is there a central traffic control centre? How well does it function? What exactly went wrong that day? What are the roles of regulation and governance, business models and budgeting, design and maintenance, training and working conditions? For every situation, we can zoom-out and ask questions, to understand the situation at hand, in its context.

Also, we can expand the solution space, which currently has only two options: pull the lever or *not*. Veer or not veer. That makes little sense, if we use this thought experiment to educate professionals: designers, engineers, managers; people like you. You can use curiosity to better understand the wider context, like we just did. You can use creativity to come up with all sorts of solutions; including very simple ones, like yell loudly to warn the workers at the track.

Moreover, if we focus on your agency, on your role as designer, engineer, or manager, you can initiate all sorts of work, on diverse domains, to prevent such accidents, like we saw above: improve design and engineering, prioritize maintenance, promote safer working conditions, collaborate with other actors to promote safety. You certainly can do more than be a bystander.

Chapter 6

Privacy is about more than 'privacy'

Often, in conversations of, for example, social media or smart cities or smart devices, people will mention the right to privacy. They may refer to the European Union's *General Data Protection Regulation*. This sometimes irritates me. Not because privacy is not important. But because there is so much more to privacy than protecting personal data. There are broader topics at stake; notably, values like justice, equality, freedom, solidarity, and conviviality.

Many people associate privacy with having control over which information about oneself, from one's private sphere, is accessible to others, in the public sphere. We can associate this view on privacy with property and transaction. In this view,

DOI: 10.1201/9781003088776-7

you own the data about yourself and can choose to give others access to these, as part of some transaction. You can choose to reveal your date of birth, in exchange for some service, which you then get 'for free'.

Some years ago, I read Maurits Martijn and Dimitri Tokmetzis' book *You do have something to hide* (2016, in Dutch). Working as investigative journalist and data journalist, they tinkered with several technologies to see how personal information is collected and monetized. They tracked the real-time brokering of their personal data as personalized ads are generated in the milliseconds it takes for a web page to load. They emulated a public Wi-Fi access point in a café in order to find out how easy one can harvest people's personal data without them noticing. I vividly remember their conclusion: as long as you do not know exactly which data are collected, by which parties, to which parties they sell your data, and for which purposes they can use these data ... you *do* have something to hide.

A wider view on privacy understands privacy as a value, with both instrumental and intrinsic value. As an instrumental value, privacy contributes to other values, such as intimacy and citizenship, for example. If you are worried of being spied upon, it is hard to develop an intimate relationship. Likewise, you need some type of privacy in order to engage in open and frank conversations with fellow citizens who may have other opinions than you.

Privacy also has intrinsic value, value in and of itself. It is an integral part of living a meaningful and fulfilling life. For one's personal development, one needs to be able to explore and express thoughts and feelings and engage in activities that you do not want to share with just everybody. Privacy is integral to autonomy. In your life, you need to be able to engage in exploration and experimentation, fully and deservedly trusting that nobody watches or assesses you.

In her book *Privacy as Power*, Carissa Véliz discusses privacy in this wider sense; in its moral, social, economic, and political sense. Poignantly, she compares data with asbestos. Both can be mined cheaply. And both can be extremely toxic. The collection and deployment of data can poison lives, institutions, and societies. Personalized ads, for example, appear harmless, but they can be weaponized to subvert the provision of news, political debates, and democratic elections. We have seen how *Cambridge Analytica* used Facebook's users' data to channel fake news to voters in order to distort elections. Véliz provides an alarming example that took place in the Netherlands during the Second World War. A Dutch Inspector of Population Registries collaborated with the Nazis to create a database of Jewish people. The Nazis used this database to find and murder many Jews. One of the things we can learn from this is not to amass too much personal data and to have ways available to delete data easily, when needed.

Central to Véliz' argument is the concept of power. States and corporations can exercise power over citizens and customers through their collection and exploitation of personal data. Moreover, they can transform one type of power into another: 'A company with economic power can use its money to gain political power through lobbying, for instance. A politically powerful person can use his power to earn money through exchanging favours with private companies'.

If we want to protect justice, equality, freedom, solidarity, and conviviality, we need to protect privacy. Privacy is not an optional feature, but a key requirement. Crucially, privacy is not an individual concern, but a collective concern, a concern on the level of society.

Let me illustrate this with the example of *SyRI* in the Netherlands. *SyRI* stands for *System for Risk Indication* and refers to legislation, administrative processes, and a technical infrastructure for the collection and analysis of citizen's personal data, using an algorithm. Dutch government agencies used *SyRI* to detect fraud with social benefits, allowances, and taxes. Several civil society interest groups brought this case to court. In 2020, the District Court of The Hague ruled that the legislation for *SyRI* is in conflict with Article 8 of the *European Convention on Human Rights*, which deals with the right to respect for private and family life. Notably, the deployment of *SyRI* involved combining personal data from tens of different databases. The court took into consideration that *SyRI* was typically applied in economically deprived areas. Effectively, people living in these areas were treated as potentially fraudulent. The court also took into account that the government chose to not disclose the algorithm's workings, in order to not enable potentially fraudulent citizens to 'game' the system. *SyRI*'s targeting practices and lack of transparency can easily lead to stigmatization and discrimination, not only of individuals, but also of groups, for example, people with certain cultural backgrounds or employment histories.

Véliz also provides a series of recommendations to curtail mass surveillance. On the level of public administration and governance, she proposes, amongst other things, to ban the trade in personal data and personalized advertising. On the level of personal habits, she proposes ways to better protect our own and other people's privacy, like 'think twice before sharing' and 'use privacy extensions and tools', and to use alterative apps, instead of the apps from tech corporations with business models based on selling personal data.

People often understand privacy as a *negative* freedom: as being *free from* other people watching or interfering. It can be useful to also understand privacy as a *positive* freedom: as being *free to* explore and express one's thoughts and feelings. We will further explore this view in Chapter 15 on virtues and flourishing, where we view technologies as tools that people can use to cultivate virtues and to create conditions for human flourishing.

Chapter 7

What is your responsibility?

As I wrote in the first chapter, I assume that you (aspire to) work in tech, maybe as a computer or data scientist, a software developer or engineer, a designer, or in a management or marketing role. In that case, you help to create the digital and online worlds that we increasingly inhabit.

In direct or indirect ways, you help to create larger or smaller systems that influence people's daily lives and the ways in which people live together. In other words, you are a moral agent.

You have responsibility for your role in the projects that you work on. Your projects produce results, for example, a prototype, a new piece of software, a product, or a system. These results enable organizations and groups of people to do things

DOI: 10.1201/9781003088776-8

and produce outcomes. And these outcomes have wider impacts on society, possibly across the globe and possibly in the long term.

You can zoom-out from your project to contemplate what type of impact you are contributing to. You can zoom-in from this impact to your project and take responsibility for your part in it.

Now, you may think: '*My project is only one small cog in a larger machine; a machine that I cannot control*' or '*I cannot singlehandedly change the business model of the company I work for*', or '*I am a scientist, an engineer, a manager; ethics is not part of my job description*'.

Of course, you are not responsible for all of your projects' outcomes and their wider impacts. But you do have *some* slice of responsibility for what you do in the projects that you work on.

In moral philosophy, responsibility is understood to be possible only if two conditions are met: knowledge and agency. You need to have knowledge about the current situation in which you may want to interfere and about future situations that may be shaped by your interference. In addition, you need to have agency (control); you need to be able to act in the current situation and help shape future situations. Without knowledge and without agency, you cannot be responsible. Crucially, the knowledge condition and the agency condition interact with each other: the more knowledge you have, the greater your responsibility for your actions, and the larger your agency, the more knowledge you will need to have in order to act responsibly.

Now, how can you take responsibility? And what can stand in the way of taking responsibility?

In *Ethics and technology*, Herman Tavani discusses several potential 'discussion stoppers', which he summarizes as four questions: *People disagree about morality, so how can we reach an agreement on moral issues? Who am I to judge others and to impose my values on them? Isn't morality a private matter? Isn't morality a matter that different cultures and groups should determine for themselves?* Let us look at each of these questions

People in many fields of knowledge disagree on correct answers to even fundamental questions, but that does not mean that they should stop having dialogues and looking for better answers. Also, based on my experiences, it can help if we distinguish between disagreements on *facts* and disagreements on *values*. In one part of a dialogue, we can focus on discussing and clarifying the facts that are available. In another part of a dialogue, we can focus on discussing and clarifying different values that different interlocutors express. (Remember Flyvbjerg's advice to combine the natural and social sciences, in Chapter 2?)

The view on ethics that is central in this book, of facilitating processes of ethical reflection, inquiry, and deliberation, is *not* about judging people. In such processes, participants are invited to express and clarify their values and concerns, their assumptions and commitments. Now, this *can* be done in a judgemental fashion, but it does not *have* to. Ideally, a meeting or workshop for ethical reflection, inquiry, and deliberation is prepared and executed with care and empathy; in ways that promote curiosity, creativity, collaboration, and compassion. Not judgement.

Furthermore, ethics is not a private matter. Ethics is concerned with living well together with others, and ethics is developed in social situations, in families, in places of learning or working. Moreover, there are different ethical perspectives, which we will discuss in the next chapters. One or more of these may have shaped your moral outlook. In these chapters, you will be equipped with a vocabulary to express your values and concerns, drawing from these ethical perspectives. Reading about these has helped me a lot, both theoretically and practically, to feel connected to and indeed supported by the people who have shaped these perspectives.

In addition, in those discussions, I hope to strike a balance between discussing agency and discussing structure. Here, agency refers to your abilities to bring about change, for example, to ask questions, to empathize with others, to collaborate, and create novel solutions. Structure then refers to the larger context in which you operate, which can impinge upon your abilities in all sorts of ways: the culture of the organization you work in can approve some perspectives or behaviours and disapprove other perspectives or behaviours; a commissioner or client can steer your project in a certain direction; or the way that the project is structured and managed, with deliverables and deadlines, can help or hinder curiosity, creativity, and collaboration.

Are you allowed to ask questions that, at first sight, appear to be outside the project's scope? Are you able to explore less obvious options, for example, the option to *not* use technology to solve a given problem, but to understand the problem differently and address it through some social intervention or social innovation? And how does the project's structure help or hinder collaboration between tasks, or between consortium partners, or to iterate between problem-setting and solution-finding?

Lastly, the question about different cultures with different ethical orientations. That is a difficult question. Answering it would require us to go into cultural and moral relativism and moral absolutism and objectivism, which I will not do here. Rather, I will give some examples of differences between cultures and ethical orientations in the following chapters. Crucially, these differences do not have to function as 'discussion stoppers'; they can be 'conversation starters'. For projects that have a global reach, for example, you will need to find ways to take into account the values and concerns of the various peoples and cultures that are involved: people whose jobs or daily lives may be affected by the project's outcomes.

We need to think globally and act locally. We need to engage with global challenges, notably by addressing the United Nation's *Sustainable Development Goals*. At the same time, we need to modify and appropriate innovations to the local contexts in which they will be used. This can be done through participation and collaboration; see Chapter 17 on Human-Centred Design, Chapter 18 on Value Sensitive Design, and Chapter 19 on Responsible Innovation.

Moreover, there *are* commonalities between cultures, especially when you, like Dr. Martin Luther King Jr., believe that 'the arc of the moral universe is long, but it bends toward justice'. Critically, this arc's bending does require that people do pull.

I do hope that we can cultivate courage to speak up when we see injustices and compassion and care to contribute to the amelioration of injustices. Ultimately, this is a book about hope: hope that the people involved in developing and applying technologies can access and mobilize their intrinsic motivations to do good and find ways to use innovation and technology to enable people to live well together.

Part II

Different ethical perspectives

In this part (Chapters 8–15), we will explore four different ethical traditions or perspectives: different ways of *doing ethics*. Each has its own distinct outlook and focus. Consequentialism looks at the positive and negative consequences of one's actions; it helps to identify and discuss pros and cons of particular technologies or solutions. Deontology, or duty ethics, looks at people's duties and rights; it helps to promote human autonomy and human dignity in relation to technologies. Relational ethics looks at people as interdependent and relational, and helps to draw attention to the effects of technologies on, for example, communication and collaboration. Virtue ethics looks at people's abilities to cultivate virtues and views technologies as tools that people can use to live well together and flourish. These four ethical traditions offer four different perspectives on the design and application of technologies. Ideally, you can combine these in your projects, according to what the project at hand requires. Each chapter is preceded by depicting a scene in which people attempt to integrate *doing ethics* in their project. These scenes do not aim to depict 'good' or a 'bad' ways of doing ethics. Rather, they are meant to depict the kind of uneasy or awkward situations that can happen when people do their best.

DOI: 10.1201/9781003088776-9

Part II

Disciplinary perspectives

Chapter 8

Software for self-driving cars

'Great that the three of you could make it on short notice', says Manfred. He is head of operations at *AutoNav*, a leading developer of software for autonomous cars. The meeting has one agenda item: the software update. Manfred has invited Lena, Stefanie, and Dieter.

They're sitting around a rectangular, white table in a spacious and otherwise empty conference room on the 12th floor. Manfred sits at the short end of the table. Behind his back are large windows through which you can see the city centre and the river that flows through it. On one long end of the table are Lena, the company's head of legal and public affairs, and Stefanie, head of user experience and design. Across them sits Dieter, head of

DOI: 10.1201/9781003088776-10

development and project manager for the update. As per usual, Dieter brought a pile of reports, which he places next to him and from which he can seize pieces of information to support what he is saying.

Dieter wants to speed-up the development of *AutoRoute*, a feature that can give their software update a competitive advantage at the upcoming trade show. They also need to look into the recommendations for autonomous driving that the European Commission published yesterday.

'I have analysed the European Commission's report'. Lena slides copies of her analysis over to her colleagues. 'They discuss various options for future policies and regulations, and give several recommendations'.

'So, they're not official?' Manfred loosens his collar with a finger. He glances at the control panel of the ventilation system, which remains puzzling to him. He looks back at his colleagues, who seem unaffected by the room's temperature.

'They're not legally binding'. Lena looks at Manfred. 'They discuss policies that *may* be developed, and possible directions in which legislation *may* go'.

'Shipping is two months from now', Dieter says dryly. 'If we move fast, we can be the first on the market with *AutoRoute*. Let's green-light this last bit of development'. He looks at Stefanie. 'Can you please talk us through last week's focus groups?'

'Sure'. Stefanie hits a button on her laptop and the projection screen switches on. 'The green bars are people who appreciate the *AutoRoute* feature. Only 30% have concerns about safety; the red bars. People also evaluated different user interface options. The almost full-screen message that pops up when you enable *AutoRoute* was most popular. It gives them a sense of control'.

Lena turns her body to Dieter, her arms crossed. 'I have not seen this graph when we met yesterday'. 'Well', he replies, 'these focus groups were informal, anyway. That's the great thing of our methodology; it allows us to show sketches before formalizing anything'.

It's five o'clock. Traffic is busy; you can see cars and lorries form long lines on both sides of the old stone bridge across the river.

'So, what's the status?' Manfred wipes his forehead with a handkerchief. 'Can we ship in two months?' 'We can', Dieter replies. He taps his index finger three times on the table top in the rhythm of his words: '... if we green-light today'.

'I have one more graph'. Stefanie clicks a button. 'In this table we compare the plusses of *AutoRoute* with our competitors' offerings'. 'We're talking to marketing about a campaign to communicate these', Dieter adds.

'Only plusses? Really? No minuses at all?' Lena inhales and looks around the table. 'I can think of a couple of risks and downsides'.

Manfred stands up and takes a couple of steps towards the ventilation control panel. He stops and turns. 'They didn't find any minuses in the focus groups, right?' 'No. I mean, yes', Stefanie points at the table on the screen, 'we only got positive feedback'.

Lena gets up and turns to look at the screen, squinting her eyes. 'You didn't *ask* about risks and drawbacks. No wonder they're not mentioning them'.

Dieter raises his eyebrows, looking at Lena, then at Manfred. 'We cannot delay the project. We need to zoom-in on what needs to be done. There is a ton of development and testing to be done'.

Lena sits down and exhales. 'Possibly. But we also need to zoom-out. We need to brainstorm broader risks and downsides. People who sleep while their cars drive them. People who make three-hour commutes because they can sleep in their car. They leave home at 5 AM while their spouse and children are asleep. And come home at 8 PM. With no energy for their family or their friends. We need to add those to the table'. She looks at Manfred.

'What if a driver puts too much trust in *AutoRoute*? You're vast asleep. The warning message flashes. I don't believe you can be awake and alert in two seconds'.

Dieter puts his right hand on his pile of documents. 'Drivers are obliged to read the Instructions and agree to the Terms and Conditions'. Stefanie chimes in, her voice a bit higher than normal. 'We can change the warning message. Font size and font colour. Or change the sound'.

Manfred notices sweat under his armpits; he tries to relax his shoulders and arms. 'Dieter, nobody wants an *AutoRoute*-accident, right?' Dieter replies, looking first at Lena and then at Manfred: 'I could make an analysis of risks and downsides'.

Manfred relaxes his arms on the table. 'Great. And Stefanie, would you be able to organize additional focus groups to discuss these risks and downsides? How long will that take?'

'We could do new focus groups in two weeks'. Stefanie feels relieved and happy to contribute.

'Great, thanks, let's do that. And Dieter, can you continue development? And could you keep an eye at the focus groups? That would be great. You're good at agile development'.

Chapter 9

Consequences and outcomes

One way to do ethics is to identify and assess the various positive and negative consequences and outcomes of actions. For each action, you can make a list of potential benefits and of potential downsides, and then choose the option that has the most or the largest benefits or the least or the smallest downsides. This approach is known as *consequentialism*. For a project you work on, you can identify its potential outcomes and impacts, and assess their benefits and downsides.

A key proponent of this approach is Jeremy Bentham (1748–1832). Bentham studied philosophy and classical languages in Oxford, worked as a lawyer, and was active in politics as a social reformer. His political and social activism included advocating

DOI: 10.1201/9781003088776-11

equality between the sexes, decriminalizing homosexuality, and championing animal rights. He set out to develop a theory for promoting wellbeing in society. In the spirit of the European Enlightenment, he based his theory on scientific knowledge, *not* on religion or on tradition. He stipulated that 'the greatest happiness of the greatest number ... is the measure of right and wrong', where he cashed out happiness in terms of pleasurable experiences ('right') and painful experiences ('wrong'). He put forward his theory as the basis for a rational system for legislation and justice. Bentham's approach is called *utilitarianism*, which refers to the *utility* or usefulness of, for example, a specific policy option; the ways in which it can contribute to maximizing people's pleasures and minimizing their pains.

This approach, of evaluating options by assessing their positive and negative consequences, is very much with us, notably in policy making and planning. A tax policy is evaluated in terms of its positive and negative effects on various groups of citizens. Or take the example of making a plan to build a bridge across a river, to develop industry and retail on the other side of the river, outside the city centre. You can put price tags on the various positive and negative consequences of that plan and make a calculation. On the minus side, you can put the costs of building the bridge, damage to various elements of the natural environment, like clean air, clean water, flora and fauna, and damage to specific parts of the local economy, for instance, to locally owned shops in the city centre, which play a role in the city's social fabric and which would lose business to the shops in the newly developed area. On the plus side you can put the monetary value of reduced commute times, the value of economic activities of the new industry and retail, which the bridge would enable. Which the city council may want to attract with subsidies, which would go to the minus side. Moreover, you can create different options for the bridge design, with different pros and cons for different elements of the environment or for different parts of the economy, and evaluate these options against each other. For example, an option that ensures relatively well-paid jobs in the new industry and retail centre. Or an option where building the bridge does less harm to the environment.

This example illustrates the difficulty of comparing apples and oranges. There are different *types* of consequences: clean air or commute times. And different consequences for different *stakeholders*: the independent shops in the city centre or the shops in the new retail centre.

It also draws attention to a key element of consequentialism: not only do you need to identify and evaluate plusses and minuses; more importantly, you will need to look at how these plusses and minuses are distributed, between people, between groups, or between countries. How are the benefits of some economic activity distributed amongst different groups of people? How are costs and benefits of global production and supply chains distributed amongst different countries? These are questions about *distributive justice*.

In consequentialism, we focus on the consequences of an act. It is based on a view on people as being able to have positive experiences and negative experiences and on a worldview of being able to identify and evaluate various potential plusses and minuses for each act.

Software for self-driving cars

Let us look at another example: self-driving cars. One can assess the benefits and downsides of a self-driving car by looking at consequences at the macro, the meso, and the micro level. On the macro level, we can look at safety and emissions. There are people who expect that self-driving cars can help to prevent accidents and to decrease emissions because they can promote safe and efficient driving. They may help reduce congestion. You can choose to leave your home at 5 AM and take a nap while your car drives you to your work, well before the rush hour. Or hop in your car at 7 PM, after the rush hour and have dinner in your car while it drives you home.

This example, however, points at negative impacts on the micro and meso levels. On the micro level, your very early or very late commute will affect your social and emotional life. You will have less opportunities to engage in sports or other activities with your friends or to enjoy time with your family and, for example, bring your child to bed. We can also look at the meso level of a city. Some expect benefits of connecting self-driving cars into a larger network, for example, in the context of a so-called smart city. The city can deploy a system that enables them to optimize the flow of traffic. An algorithm may grant priority to emergency services' vehicles, like police cars, ambulances, and fire trucks, so that other vehicles automatically swerve and make room for them. Such an option can, however, also be used in controversial ways. You may be able to buy an expensive car that includes a service to request and receive such priority. That would be nice for those who can afford it. Remember New York City planner Robert Moses' low-hanging bridges that prevented people who rely on public transport to travel to the beaches?

Another controversial issue would be data ownership. The idea of a 'smart city' assumes the collection of large amounts of data, which are used to make predictions and control traffic. But who owns these data? The city council? The citizens? Some transnational tech company? A public-private partnership? Which public organizations or private companies exactly? And which organizations or companies will they share the data with? For what purposes and under which conditions? Addressing such questions will determine which outcomes we include or exclude (possibly unknowingly) in our analysis and whether we evaluate these as positive or as negative.

Limitations and critique

Despite its widespread usage and intuitive appeal, consequentialism has several limitations. The utilitarianism of Bentham, for example, is based on the assumption that we are able to assess the pleasures and pains that follow from any action. Bentham even developed a *hedonistic calculus*: a large table that lists hundreds of types of experiences that people can have, like eating food, with instructions to calculate the pleasures and pains for each possible experience, based on their intensity, duration, (un)certainty, closeness or remoteness in time, fecundity, purity, and extent. He wanted to put numeric values on all human experiences. Although interesting in theory, the evaluation of all these pleasures and pains will be hard in practice, since many of our experiences are subjective and qualitative, rather than objective and quantitative.

John Stuart Mill (1806–1873), who was educated by Bentham (a good friend of his father, James Mill), criticized utilitarianism for its inability to distinguish between different *qualities* of pleasures. Mill advocated differentiating between higher quality pleasures and lower quality pleasures. Attentively reading great poetry will give you a high-quality pleasure. Mindlessly playing a silly game will give you a low-quality pleasure. In addition, Mill's take on utilitarianism stresses freedom: the freedom to develop to your full potential, not restricted by other people; as long as your freedom does not interfere with other people's freedoms. Mill was a Member of Parliament and a proponent of women's emancipation and the abolition of slavery.

Moreover, it will be difficult to put different plusses and minuses into one sum. Like comparing apples and oranges. In the example of a self-driving car, the safety of the driver will be a plus. The safety of pedestrians, however, may be a minus because they typically remain vulnerable against cars, especially if the driver pays little paying attention to what happens outside the car. Moreover, if we focus on all sorts of new functionalities that become possible by collecting drivers' data, we may have plusses for better real-time travel advice and cheaper, tailor-made insurance policies; and minuses for infringements of privacy that the collection and utilization of drivers' data typically entails and risks of discrimination and diminished solidarity that such personalized insurance policies may entail. It is hard to put such diverse issues into one sum. It is beyond comparing apples and oranges. More like comparing bananas and broccoli, soy and sauerkraut.

When made into a caricature, consequentialism can even become horrifying. There is the infamous and hypothetical example of the five patients in a hospital. Each is seriously impaired and in need of one vital organ. The surgeon will be able to save all five, if only he had five healthy donor organs.

You have a minor injury and enter the hospital. Taking consequentialism to an absurd extreme, the surgeon may want to sacrifice you in order to save these five. Obviously, this is not a society we want to live in. This is not a calculation we want to make. (If, however, you would want to make a calculation, you would need to take into account the downside of everybody living in constant fear of being sacrificed in scenarios like this.)

Practical applications

Another infamous example is the *Ford Pinto*. In 1978, a truck rear-ended a five-year-old *Ford Pinto*, with three young women in it. The collision caused the petrol tank to explode, which killed all three. The Elkhart County jury returned a criminal homicide charge against the Ford Company. During the following trial, investigations brought to light that Ford engineers had been working with a novel design, to make the car small and cheap, and had placed the petrol tank just behind the car's rear axle and gears. In a rear collision, the rear axle and gears would puncture the petrol tank. Ford knew about this risk and had made a cost-benefit analysis: retrofitting all vehicles with this shortcoming would cost $137 million (12.4 million vehicles × $11), whereas accepting this shortcoming and paying insurance claims would cost less in dollars: $50 million (180 lives lost × $200,000, plus 180 people seriously wounded × $67,000, plus 2100 cars destroyed × $700). Strikingly, the jury found Ford *not* guilty of criminal homicide. Ford nevertheless chose to recall 1.5 million vehicles for retrofitting and they had to pay millions of dollars in legal settlements to accident victims.

Assigning a monetary value to a person's life may strike you as misguided or evil. Maybe it is. Maybe it sometimes needs to be done; if only as a theoretical exercise. In the United Kingdom, the *National Institute for Health and Care Excellence* has been using QALY as a metric to estimate the effects of various medical treatments on people's quality of life. QALY stands for *Quality Adjusted Life Year*; one year in perfect health gets 1 QALY. One year of partial healthy living, because of some disability or medical condition, gets a value between 0 and 1. You can imagine the difficulties of attributing a specific value to a specific condition. A missing leg. A missing arm. Difficulty breathing. Having constant pain. Having lungs that work for 50%. Or a heart that works for 50%. Would that count as 0.5? The goal of using QALYs is to make informed decisions about the allocation of scarce resources. For a specific medical treatment, one can assess, on the one hand, its costs and, on the other hand, its contribution to improving QALYs. Such calculations introduce a range of problems. For example, what is the value of increasing a QALY from 0.2 to 0.5, from suffering to a moderate quality of life, compared to increasing a QALY from 0.5 to 0.8, from a moderate quality of life to good health? Quantitatively, these increases are equal (0.3). Should we give priority to people who suffer most, those who start at 0.2? And what if, in the latter example, the QALY would increase from 0.5 to 0.9? Should we prioritize a 0.4 increase (0.5 to 0.9) over a 0.3 increase (0.2 to 0.5)?

Numbers and metrics

Consequentialism is often associated with numbers and metrics. In the example of the *Moral Machine* survey (Chapter 5), people counted and compared casualties. This example illustrated the challenge of weighing the lives of pedestrians, who are strangers to you, against passengers, people whom you know, or the life of a medical doctor against the life of a homeless person.

Maybe here is a good place to make some remarks about different types of information: quantitative and qualitative, and objective and subjective. I have found myself in situations in which I talked about something that I value, say, physical or mental health of people, and in which people asked me to *quantify* that concern; to put numbers on this phenomenon. As if, without numbers, a concern is less valid. In contrast, I once heard, in an interview about traffic casualties, a scientist give exact numbers. The interviewer then asked the scientist to *qualify* these numbers. To evaluate these numbers. Are these high or low? The scientist was slow to answer. I hope to hear more questions about qualifying numbers; more talk about what these numbers mean. Now, concerning the difference between *objective* and *subjective* information. With a bit of glossing-over, I would propose that *objective* information is independent of people's interpretation (or *intersubjective*, when different people look at the same phenomenon with the same method and have similar perceptions and interpretations), whereas *subjective* information does involve people's personal perception and interpretation.

We can now create a 2×2 matrix, with four types of information. An expression of *quantitative* and *objective* information would be: 'the length of this plank is 1 meter'. When I fill-in a 'subjective wellbeing survey', I generate *quantitative* data about my *subjective* experience. Examples of *qualitative* and *objective* information could be the various rights in the *European Convention on Human Rights*; they do not deal with numbers, but they are articulated in *objective* terms (or *intersubjective*, because interpretation can require discussion). Lastly, expressing one's values can be an example of information that is *qualitative* and *subjective*; hard to put numbers on and dealing with one's personal experience. The objective of this summing-up is to convey the ideas that there are different types of information (not only quantitative and objective data) and that all of them are potentially at play and relevant in your project.

It is also worthwhile to look at the challenge of finding appropriate metrics; what some people refer to as *Key Performance Indicators*. For mobile apps, for example, 'time spent in app' is often used as a key metric. For apps that are based on capturing and monetizing people's attention, that is. Dating apps provide an interesting case. Some of these are designed to maximize 'time spent in app' or 'swipes per week'. Now imagine such an app with a different metric: 'the quality of the time spent with the other person in real life'. This may entail a different business model, probably based on subscriptions, not on advertising. We saw the problem of choosing appropriate metrics also in our discussion of GDP and the need to go 'beyond GDP'.

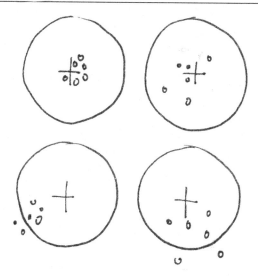

In addition, it is worthwhile to discuss *accuracy* and *precision*. Imagine that you have a bow and are shooting arrows at a target. As a metaphor for using an algorithm to identify, for example, instances of suspicious or fraudulent behaviour. You want the algorithm to point at fraudulent behaviours (true positives) and to avoid false positives (pointing at what later turns out to be not-fraud) and false negatives (not-pointing at fraud). Now imagine that there are different algorithms to shoot the arrows: algorithms with more or less *accuracy* (or *bias*), which refers to how close the arrows' hits are to the target, the bull's eye, and algorithms with more or less *precision* (or the inverse of *variance* or *noise*), which refers to the spread of the arrows' hits. This gives four different algorithms: *accurate and precise*, where arrows are close to the centre, in a small cluster; *accurate and imprecise*, where the arrows are around the centre, but scattered; *inaccurate and precise*, where arrows are away from the centre, in a small cluster; and *inaccurate and imprecise*, where arrows are away from the centre, and also scattered.

Anyone will prefer the first one: an *accurate and precise* algorithm. But this will probably not be available, so that we have to negotiate between accuracy and precision. Which would you prefer? I would prefer accuracy over precision. I would choose an *accurate and imprecise* algorithm, with many arrows close enough to the bull's eye, and accept some imprecision. This would involve, for example, using data that are accurate in that they approximate the phenomenon that I am interested in, even if they are less imprecise, for example, because they are more qualitative and less quantitative. In some projects, however, I have noticed that people can tend to prefer precision over accuracy. They prefer precise, quantitative data, even if these numbers less accurately measure the phenomenon that they are interested in. The arrows are in a small cluster, but away from the centre. I would argue that such a choice gives an illusion of precision. Surely, I am not against metrics. But I would avoid numbers fetishism.

You may recognize these issues in the following examples. You may have found yourself filling-in some questionnaire, with tens of precise questions and very precise answer options, maybe a nine-point Likert-scale. But after ploughing through these questions, you may wonder whether you were able to accurately express your concerns, what matters to you. Maybe you would have preferred a less-precise, more-accurate way of expressing your concerns? Or, in another example, of an online server that logs people's behaviours and produces precise data. You can follow each individual user on the level of clicks and scrolls. You can describe precisely what they do. But can you accurately understand their experiences? Are they clicking multiple times from A to B to C because they are interested in A, B, and C? Or are they lost and irritated and do they hate having to go from A to B to C and are they instead interested in P, Q, and R?

System boundaries

Another challenge of consequentialism concerns the determining of system boundaries. Which types of consequences do you include or exclude in your analysis? Which people or groups do you include or exclude? Only those who are immediately affected? The owner of the self-driving car. Or also those further removed? Pedestrians? We may tend to include those factors that can be quantified relatively easily and to leave qualitative factors out of the analysis. And what about car makers who prioritize the car owners' concerns over pedestrians' concerns, or city planners who prioritize their own prestige by procuring some shiny, new 'smart' system?

Moreover, there is a time dimension to determining system boundaries. Which time frame will you use to assess potential pros and cons? Of course, you will look at immediate effects on the short term. But what about less direct, second-order, or longer-term effects? And if you want to include the latter, how can you assess these? When economists talk about what falls inside or outside their analysis, they talk about *externalities*: the effects that they choose to ignore. Sadly, such externalities typically relate to costs. That is why many supply chains start on the other side of the globe, in a rare-earth metals mine in a conflict zone, or in a sweatshop where people labour in unhealthy and unsafe circumstances, and why many products are discarded and end up in toxic dumps in distant countries. Out of our sight.

Had the externalities of damage to the planet and harms to people been taken into account, then this chain of production, consumption, and disposal would look very different. Not necessarily more costly. It is possible to have a circular economy, organic food production, renewable energy, fair trade arrangements, and better labour conditions in ways that are cost effective and commercially viable. What we need is legislation that puts price tags on these externalities, so that they are taken into account, above the bottom line.

Let me share an example from moral philosopher and activist Peter Singer (born 1946), to illustrate the need to zoom-out and to expand your moral circle, your circle of concern. Singer invites you to imagine a shallow pond and a child falling into it, almost drowning. You can choose to step into the pond and save the child without any danger for yourself. You will, however, ruin your new pair of shoes. Most people say they will save the child. Singer then asks why we let millions of children die of malnutrition or of preventable diseases in remote countries. Do we save the child in the pond because she is visible and close-by? Do we let other children suffer because they are invisible and far-away? We need to zoom-out and look at consequences that happen outside our immediate perception; removed in space and time. This is especially pertinent with regards to the climate crisis, to take into account the needs and interests of people in the Global South and of future generations, and to strive towards distributive and intergenerational justice.

Other traditions

Finally, and as bridges to the next chapters, we can look at how other ethical traditions can help to remedy some shortcomings of consequentialism. One such shortcoming is that it can lead one to choose to realize benign consequences through less-benign means, so that 'the end justifies the means' (as in the example of the surgeon). Rather than looking only at consequences, we need to also look at duties and rights; we will do that in Chapter 11.

Another shortcoming concerns the assumption that we need to look objectively and impartially at the world and not be distracted by specifics. Conversely, one can argue that we *do* need to look at the specifics of each situation. Notably, we would need to take into account the various relationships between people. You may, for example, want to treat your child, sibling, parent, or friend differently than you would treat a random person. In Chapter 13, we will explore the role of relationships.

In my experience, consequentialist analyses tend to deal with comparing options. You assess plusses and minuses for options A and B and conclude that A is better than B. You may then forget, however, to question why we would want A or B in the first place. To what further end are A and B potential means? Also, there may be an option C that will help to work towards that end much better than A or B would. You may want to ask questions like: *What type of society do we want to contribute to with this project? How can we enable people to use technologies as tools to cultivate virtues that they need in order to live well together?* We will address such questions in Chapter 15.

Take a project that you work on. Imagine that this project goes well, delivers results, and has impacts in the world: in society, in people's daily lives. Now, take a piece of paper and draw a vertical line. Then make two lists: on the left you write potential positive outcomes; on the right you write potential negative outcomes. Consider different domains, for example, meaningful employment and sustainable development, bodily and mental health, or pollution and emissions. Consider different groups of people, who will experience different benefits or downsides. Also explore potential, unintended, undesirable outcomes. See whether you need to extend your analysis' system boundaries. Now, have a look at your lists. How do you feel about these? Are there ways to mitigate some of the negative impacts? Or improve some of the positive impacts? What are critical benefits? What are critical downsides? Can you weigh them against each other? How can you proceed with your analysis?

Chapter 10

Cameras in public spaces

'Thanks Hugo, that was a very clear presentation'. Lucia looks around the table, at Carmen and Francisco. Hugo returns to his seat. They sit around an oval table made of dark oak wood, in the middle of the main room of the city hall, from the 16th century. The regular conference rooms were booked so they had to convene here. Normally, this room is used for formal meetings.

Lucia has served as a city councillor for five years. She is popular and she wants to continue to make a positive impact on the city and its inhabitants. Last year, she met Hugo, CEO of *UrbanX*. That was the start of making plans to install cameras in the city centre and to use machine learning to promote

DOI: 10.1201/9781003088776-12

safety. The objective of today's meeting is to structure these plans into a project.

Carmen is the project manager. She understands both the technology and the administrative and political processes, and she has a track record in managing public-private partnerships. She sees this project as a springboard to larger projects, possibly abroad. Across the table sits Francisco, who reports to her. He has worked hard to prepare for this meeting. He has been looking into ways to make the new system interface with existing hardware and software.

'Here's a brochure of projects for other cities that we have worked with'. Hugo points at a pile of colourful flyers in the middle of the table. 'Based on what Francisco and I have been discussing so far, I believe that the *ML4* would be most suitable'. 'Indeed. I have prepared several tests to make our systems compatible with the *ML4*', Francisco adds.

Carmen moves her body in her chair and straightens her back. 'Great. Let's discuss our planning. We have two weeks to make a proper project plan. This is our window of opportunity to get it into next year's budgeting cycle'.

Lucia's eyes roam the wall opposite her. It has wooden panelling from floor to ceiling. And a grid of some 40 life-sized oil paints of former mayors. Solemn faces, staring at them. 'Good point, Carmen. However, I'm not entirely happy about security and privacy'.

Hugo looks at Francisco and then at Lucia. 'The *ML4* is fully compliant with current legislation for data security and privacy protection. Right, Francisco?'

'I meant *people's* security and privacy; *not* of the data', Lucia interrupts him. 'Carmen, could you please talk us through how these cameras and software will *not* stigmatize and discriminate?'

Carmen feels some tension in her neck and shoulders. She looks at Hugo. 'This citizens' rights group did bring concerns to our attention. Some citizens have serious concerns about systems like this, about the risks of discrimination. Obviously, we all want to prevent issues. We want a system that does not discriminate'. 'And we need to respond to this group', Lucia adds.

Carmen nods. 'I'm sure we can find ways to prevent these issues with the *ML4* modules. I mean, we will be able to protect people's safety, and at the same time protect their privacy'.

Hugo opens the palms of both his hands and gestures at the pile of flyers. 'We have a solid track record in this subject matter. The *ML4* can be modified to minimize the risks of discrimination. Our tech is state of the art. The field has come a long way since the days that face recognition labelled people of colour as gorillas'. Lucia raises one eyebrow. 'I am sorry. What did you say? It *minimizes* the risks of discrimination? So it *does* do that; it *does* discriminate?'

Francisco takes this as a cue. 'Well, in a way, *discriminating* is exactly what any machine learning system does. It puts different cases in different boxes, it puts people in categories. This person A is likely to behave violently. This other person B is not. And there will be errors. False positives and false negatives. And you can choose which type of errors you want to minimize. That's what we mean with *optimize*'.

Lucia turns to him. 'Okay, great. And how does that relate to citizens' rights? I believe this is not a math problem. The people in our city have rights, and they are raising concerns and pushing back. We need to do this very carefully'.

Hugo looks around the table and is about to speak. Lucia interrupts him. 'This not about optimizing. It's about keeping citizens safe. Not about selling their data to some private company. I mean, Hugo, I have absolutely no reason to mistrust you and your company. But the thing is, there have been too many examples of collecting and selling data. Maybe not necessarily illegal. But still. Hard to explain to a journalist in a press meeting'.

Another cue for Francisco. 'We've worked hard to make sure everything works fine: both technology-wise and legal-wise'.

From outside the building, loud sounds of construction work enter the room. There is always some construction work going on in the city centre. The rumbling sound is like the sound that a huge pile of bricks would make when they are being poured on slabs of concrete. It goes on for a good couple of seconds.

Lucia glances at her watch. 'My real concerns are justice and equality. Face recognition does not work perfectly, I do understand that. Moreover, the police need a transparent and legitimate process of gathering evidence for cases that they want to investigate or prosecute'.

Hugo nods. 'I am entirely on board with your concerns, Lucia. We've always held these values to guide our product development. That's why so many other cities have chosen our systems'.

Carmen looks at Lucia. 'We will make sure to take into account your concerns for justice and equality. We will indeed need to work with utmost care'.

Then she looks Francisco. 'Can you please walk us through the measures that you and Hugo propose that we will need to put into place in order to make the systems compatible? After that, I guess we can discuss what needs to go into the project plan'.

Chapter 11

Duties and rights

Alternatively, you can think about ethics in terms of duties: the duties that you have towards other people, and possibly also towards yourself, towards non-human animals, and towards the natural environment. This approach is called duty ethics or *deontology; deon* is Greek for duty.

Rather than looking at your actions' consequences, you identify *duties* that you will need to fulfil. For a project that you work on, you can reflect on the duties that you need to fulfil while working in the project. In addition, you can envision your project delivering results, for example, a new system or application, and reflect on how these can help or hinder

DOI: 10.1201/9781003088776-13

people to fulfil their duties. (Note that this example mixes duty ethics with a bit of consequentialism).

In duty ethics, the moral evaluation of an action is based on whether that action is in agreement with a moral duty, independent of that action's consequences. Imagine that you walk in the city centre with a friend and that you meet a person who begs for money. If you give money out of your will to follow some moral duty that you feel inside, for example, a duty to give to the poor, you would do the right thing, according to duty ethics. If, however, you give money in order to create some positive effect, for example, to make a favourable impression on your friend or to give yourself a warm glow, then you would *not* necessarily be doing the right thing according to duty ethics. (In contrast, a consequentialist view would focus on the outcomes of giving the money and on the consequent actions of the recipient, how they spend the money, for example.)

Legal duties and moral duties

Across centuries and across cultures, different peoples have articulated different duties. Interestingly, there is some overlap between *moral* duties and *legal* duties. In the *Ten Commandments*, which play a key role in Judaism and Christianity, we can recognize *moral* duties, such as 'honour thy father and thy mother' and 'thou shalt not commit adultery', which function like moral norms and which can differ between groups or cultures; and *legal* duties, such as 'thou shalt not kill' and 'thou shalt not steal', which are codified in laws and which differ much less between cultures. Moreover, there are duties that can be moral or legal, depending on the context. Telling the truth, for example, is a moral duty in many informal contexts; you are expected not to lie, but you can tell a white lie or an incomplete truth if that is socially appropriate or acceptable. In formal contexts, however, in writing and signing contracts, for example, you have a legal duty to act truthfully.

It is worth noting that moral norms can change over time and differ between cultures, and legal norms can change over time and differ between jurisdictions. When Rosa Parks, an African American woman, sat on a seat in the 'colored section' of a public transport bus in Montgomery, Alabama and refused to give up her seat, she did something that was considered illegal. In 1955, the state of Alabama had segregation laws. The *National Association for the Advancement of Colored People* took this case to court, as part of their civil rights activism, which eventually led to changes in the US legislation. Rosa Parks' act was first viewed as illegal, but was later found to be morally right and has led to more just legislation.

There are also examples of the opposite: acts that are legal but not morally right. Think of corporate lawyers who find loopholes in tax laws, or business managers who exclude so-called 'externalities' from their spreadsheets to boost their profit margins. This may be legal, but it is increasingly viewed as morally wrong. There are many examples that involve extraction and exploitation: corporations that extract value from the society in which they operate, for example, when they use infrastructure or research outcomes that were paid for by the public through taxes; companies that exploit their employees, paying so little that they need to work multiple jobs, or exploit gig workers, who have very few labour rights; or firms that extract value from nature and destroy natural systems, for example, taking fresh water from a river and secreting polluted water back into it.

Now, most innovations are legal, in the sense of 'not illegal'. We may, however, find some innovations morally deficient, for example, when companies do not account for harms they inflict on people or on nature. Worse, companies can dodge legislation and choose to risk fines, when their illegal operations' benefits outweigh paying these fines. (You may recognize the latter example as a corrupted consequentialist way of reasoning.)

Universal duties

It can be frustrating that duties differ across cultures and that they can change over time. Surely, there must be *universal* duties; duties that are true in all contexts and in all conditions? This was a question that Immanuel Kant (1724–1804) struggled with. Kant studied and taught theology, philosophy, mathematics, and physics in Königsberg (present-day Kalingrad). Relatively late in his career, he wrote a series of immensely influential books, several of which dealt with ethics.

Kant, a key proponent of the European Enlightenment, wanted to find moral principles *not* in religion or tradition, *not* in the demands of priests or of kings, but in *reason*. Kant came up with a *categorical imperative*, a universal rule or 'maxim': 'Act only according to that maxim whereby you can, at the same time, will that it should become a universal law'. You articulate a rule that you can will to be a universal rule and you then follow this rule. Strikingly, Kant did not add any normative content or direction. Instead, logical reasoning alone would do the trick.

You cannot logically will a rule like '*It is okay to steal*', because such a rule is not logical. The act of stealing contradicts the idea of property because the thief denies the owner's property. At the same time, however, the act of stealing affirms the idea of property because the thief's objective is to own the object that he stole. Similarly, a rule like '*It is okay to break promises*' is logically impossible because it contradicts the practice of making and keeping promises. If you cannot trust others to keep their promises, the whole idea of making promises collapses.

Kant provided an alternative formulation for his categorical imperative: 'Act in such a way that you treat humanity, whether in your own person or in the person of any other, never merely as a means to an end, but always at the same time as an end'. This sounds more practical. Suppose that you work on the development of some social media app. Do you treat the person who uses this app 'merely as a means' to collect data, which the company can sell to whoever pays for it? Or do you treat these people with respect for their dignity and autonomy, so that using the app can empower them in pursuing their own ends?

Kant's ethics is based on 'reverence for the moral law', a law that is supposedly within each rational person and that depends on that person's 'good will', their good intentions. Obeying a duty can require of you that you go against your nature or your impulses or against customs or norms of other people.

In addition to duties, we can articulate corresponding rights, so that, for example, the duty of person A to treat person B in a certain way corresponds with the right of B to be treated in that way by A. This can be backed up by a legal system in which the duty of A to stick to a certain agreement with B corresponds with B's rights to have A to stick that agreement. If this is a legal right, then B can demand enforcement or reparation through a judicial process.

In deontology, we look at duties and rights. We view people as rational beings and focus on human dignity and human autonomy. It assumes our ability to articulate and fulfil duties and to articulate and protect rights. Deontology starts from a moral agent's good will.

The influence of duty ethics can be found in *Codes of Conduct* or *Codes of Ethics* of companies or organizations. The *Code of Ethics* of the *Association for Computing Machinery* (ACM), for example, codifies a range of ethical principles (to contribute to society and to human wellbeing; to avoid harm; to be honest and trustworthy; to be fair and take action not to discriminate; to respect privacy and honour confidentiality), professional responsibilities (for example, to strive for high quality and perform work only in areas of competence), and leadership principles (for example, to manage people in ways that enhance their quality of life and facilitate their growth).

Limitations and critique

A key critique on deontology is that *universal* duties will not always work in *specific* contexts and will need to be adapted. Notoriously, Kant held that one ought to always speak the truth, also in the example where your friend comes to hide in your house in order to flee a person who intends to kill them. When this potential murderer knocks at your door and asks whether your friend is in your house, you will need to speak the truth, according to Kant. I guess that most of us would find it more appropriate to lie in this situation.

Another limitation of deontology is that it offers no obvious ways to prioritize or choose between conflicting duties. When you work in a company, you can think of a conflict between a duty to work on a project even though you have come to realize its negative effects on society or on nature, as well as a duty the duty to speak up, in order to protect your fellow citizens and nature.

Whistle blowers are typically challenged by such conflicting duties. In 2013, Edward Snowden, former employee of the US Central Intelligence Agency and former contractor for the US government, copied and leaked classified information from the National Security Agency. He had to combine his duties as a civil servant and his duties as a concerned citizen.

Furthermore, it is not immediately clear, in duty ethics, how you can account for the specifics of a practical situation and, for example, the relationships between specific people. You may recall a significant conversation in one of your projects, with a person who matters to you, about an issue that matters to you. Possibly, you had to balance various duties: to speak candidly about your concerns, to support that person in what they value, to look at the bigger picture of the project, to protect others from potential harm caused by your project's outcomes, to get your work done in time, and to be home on time to spend time with your family. All these duties matter. You will need to take into account your *relationships* with others and the specific situation you are in. We will come back to that in Chapters 13 (relational ethics) and 15 (virtue ethics).

Practical applications

Duty ethics, and its emphasis on human dignity and human autonomy, has had a big influence on the creation of human rights. I use the word 'creation' deliberately. Human rights did not fall from the sky. Or grow on trees. Rather, they were created by people, over the course of centuries. One may point at ancient attempts to articulate human rights, like the *Code of Hammurabi*, a code of law of ancient Mesopotamia from around 1750 BCE, the *Edicts of Ashoka,* a series of inscriptions in stone in public spaces, attributed to Emperor Ashoka, who reigned between 268 and 232 BCE, or the 1215 *Magna Carta*, a charter of King John of England that gave rights to barons (but not to common people).

Only fairly recently have human rights been codified in international treaties and in national laws. It is important to understand the relationships between people's interests and rights. John Tasioulas, professor of ethics and legal philosophy and director of the Institute for Ethics in AI at University of Oxford, has argued that only those interests of people that lead to duties of others can be understood as rights. Imagine person A who suffers from a kidney disease and, therefore, has an interest in receiving a donor's kidney. Most people would agree that nobody has a *duty* to donate their kidney to A. (Though some may *feel* such a duty.) So, A does not have a right to receive a kidney. Now imagine person B, who has an interest in not being discriminated against. This interest *does* translate into duties of other people and organizations *not* to discriminate B. This has been codified in laws that prohibit discrimination, so that B does have a right not to be discriminated against.

One starting point to understanding human rights is the (non-binding but authoritative) *Universal Declaration of Human Rights*, issued by the United Nations' General Assembly in 1948. It has provided the basis for a series of binding international treaties, notably the *International Covenant on Civil and Political Rights* (in force since 1976; signed and ratified by over 170 countries; signed but not ratified by several, including China), which covers rights to life, freedom of religion, of speech, of assembly, and of movement, electoral rights, and rights to due process and a fair trial; and the *International Covenant on Economic, Social and Cultural Rights* (in force since 1976; signed and ratified by over 170 countries; signed but not ratified by several, including the US), which covers areas like education, food, housing, health care, and employment. These treaties were followed by a series of more specific *Conventions*: against racial discrimination (1969); against discrimination of women (1981); for rights of children (1990); and for rights of people with disabilities (2006). These treaties are part of *international public law*, which means that they are binding on the level of nations (and international organizations) and arguably create obligations also for other actors, such as corporations and individuals.

In addition, concerns for human rights have been adopted in many nations' public or constitutional laws; these codify the duties of a government towards their citizens and the rights of citizens towards their government. Moreover, there are treaties that enable citizens to bring cases against their government to court. Notably, citizens of states that are part of the *Council of Europe* (an international organization with 46 member states; not to be confused with the *European Union*, a supranational organization with 27 member states) can invoke the *European Convention on Human Rights* and bring cases to the *European Court of Human Rights* in Strasbourg, France.

This is what several civic organizations and citizens did in the Netherlands. They successfully asked the court to prohibit the Dutch government from utilizing *SyRI* (see Chapter 6). Somewhat similarly, citizens of the 53 states on the African continent that ratified the *African Charter on Human and Peoples' Rights* and citizens of the 24 states across North and South America that ratified the *American Convention on Human Rights* can bring cases to, respectively, the *African Commission on Human and Peoples' Rights* and the *Inter-American Commission on Human Rights*; and in certain cases, these disputes can be brought to, respectively, the *African Court on Human and Peoples' Rights* and the *Inter-American Court of Human Rights*.

The legislations discussed so far dealt with protecting citizens from states' abuses or misuses of power. Now, how are human rights relevant to the design and application of technologies? How are human rights relevant for companies and the innovations they work on? More to the point: How are human rights relevant for your project? Maybe you feel that you need to steer your project, voluntarily, in a direction that is more aligned with concerns for human rights? If you work for a government body or associated agencies, you are typically required to respect human rights and relevant legislation. In addition and crucially, human rights refer *not only* to legislation.

The notion of human rights can help to design non-legal policies or measures that can help to respect, protect, and fulfil human rights, notably rights that pertain to human dignity and human autonomy. In your projects, you can work to respect, protect, and help to fulfil human rights, without the need to fully understand all sorts of international treaties and national laws.

Rather accessible guidelines and recommendations have been issued to support this. Some of them focus on promoting human rights in the design and application of algorithms and AI. The European Commission's *High Level Expert Group on AI*, for example, based their *Ethics Guidelines for Trustworthy AI* on the *European Convention on Human Rights*. Furthermore, the United Nations' *Office of the High Commissioner for Human Rights* (headquarters in Geneva, Switzerland) and the *Organisation for Economic Co-operation and Development* (headquarters in Paris, France) have initiatives to better align businesses and concerns for human rights.

Cameras in public spaces

Let us look at a practical example of duties and rights. Numerous cities have cameras in public spaces. Increasingly, these systems include image recognition functionalities, so they can recognize vehicles' license plates and people's faces. We can imagine that a city council, typically in collaboration with one or more companies, in a public-private partnership, chooses to install cameras in order to monitor people's behaviours out of their duties to protect citizens against harm and to combat crime. Cameras, equipped with microphones and other sensors, can register, for example, a fight and send an alarm to the police, so they can go there and protect citizens. During large events, cameras can capture images that can support first-responders in their work.

This duty to promote citizens' safety needs to go hand in hand with a duty to respect, protect, and fulfil citizens' rights to privacy. In such a context, safety and privacy are often framed as opposites. However, they do not necessarily have to conflict; one can envision systems that uphold *both* safety *and* privacy. Similarly to how a tent can be *both* spacious *and* lightweight, if it is made of lightweight, waterproof cloth and uses flexible poles to envelop a large volume.

Moreover, with regards to the design and application of technologies in collaboration with government bodies or associated agencies, you will need to think carefully about *legitimacy*, whether the goals the system aims to realize are legitimate; *effectiveness*, whether this system can actually help to realize these goals; and *proportionality*, whether the harms, for example, the infringement on people's privacy, are proportional to the benefits the system aims to offer, for example, to prevent harm and combat crime. The Dutch *SyRI* system did not pass the effectiveness and proportionality criteria. For the case of cameras in public spaces, you will also need to consider, for example, the risks of undermining people's rights to freedom, expression, and assembly. Face recognition can deliver lists of people who meet in public spaces. Speech recognition can deliver transcripts of their conversations. Such technologies can have chilling effects on people's liberties and corrode public and civic values, like freedom, democracy, and conviviality.

Fairness and 'errors'

Moreover, technologies that rely on machine learning can perpetuate or exacerbate existing biases. Machine learning algorithms process data from past events to find patterns and make predictions for future events. Give the software a stack of photos with labels, for example, for cats and dogs and it can 'learn' to put these labels also on new photos. This may sound 'objective' and 'neutral'. But very often it is not. A notorious incident happened with Google's image recognition in 2015. Two black teenagers found that a picture of them was labelled as 'gorillas' by Google. Very likely, the software had been trained with mainly photos of people with light skins and very few photos of people with dark skins. The system's best guess for people with dark coloured skins was 'gorillas'. Google apologised. In order to fix this problem, they could have trained their model with a more diverse set of photos. Strikingly, however, Google 'solved' this problem by removing several potentially problematic labels, like *gorilla, chimp, chimpanzee,* and *monkey.* A rather 'quick and dirty' fix, one may argue.

Sadly, this example is not unique. In her book *Algorithms of Oppression,* Safiya Umoja Noble, discusses many more harmful examples of such stigmatization and discrimination. Similarly, in *Weapons of Math Destruction,* Cathy O'Neil discusses how the design and application of algorithms can exacerbate existing unfairness and injustice. It reads like a catalogue of harms that algorithms can bring about in all domains of society and of daily life: in education and exams, in employment and careers, in loans and insurances, in health care and social services, in policing and judiciary processes, and in online social networking and dating. Moreover, these harms often affect people who already have less opportunities, money, or power.

Some people would look at these examples and say that these algorithms are not unfair or unjust, but that they make 'errors' and that they can be fixed. In the vocabulary of data science, such 'errors' are typically understood as *false positives* or *false negatives.* Certainly, such errors need to be critiqued, investigated, and corrected. But that is not enough. That is why there are inverted commas around 'errors'. Very often, algorithms' errors point at *systemic* unfairness or injustice. The scope of critique, investigation, and correction needs to be broadened. It is not about fixing errors in the algorithm; it is about changing large systems. An infamous and well-known example is *COMPAS,* an algorithm that judges can use to assess the likelihood of recidivism. This algorithm was found to be racially biased: it gave black defendants higher scores for recidivism than white defendants. Of course, this bias needs to be corrected. But it also points at the need for a wider critique, investigation, and correction of systemic inequality and unfairness.

Take a project that you work on: developing a new system or application. Imagine that your project goes well and that it delivers functionalities. Which are these functionalities? For each, explore whether there are duties that can be expected from you in your work in delivering these functionalities. Similarly, explore whether there are rights associated to these functionalities. How can you respect, protect, and help to fulfil these rights? Are there codes of conduct or standards that you need to comply to? You can go one step further. You can look at the duties and rights of the people who will use this system, and the duties and rights of the people who will be affected by the application of this system. You can do a 'what-if' exercise, loosely based on Kant's categorical imperative. Imagine that everybody, everywhere, all the time, uses these functionalities. What would society then look like? How will people spend their time and interact with others? What would economic, social, cultural, and political processes look like? Do you find such a future worthwhile to contribute to and work towards?

Chapter 12

Smart devices in our homes

'How do we want to relate to the various technologies that we interact with in daily life?' Mikael looks into the camera. Then the camera pans to bring into view the three panellists. Mikael went from journalism to event organizing several years ago. He hand-picked the three participants.

Per, a spokesperson of the company that introduced the *HiperSpeaker*: a white cylinder that uses cameras to sense your mood and plays different types of music to match your different moods. Sofia, an artist-slash-developer, who created the *HypeSeeker*, not only as a parody of this particular device, but also to spark a wider discussion on such devices. And Annika, an academic and citizens' rights activist. The four are sitting

DOI: 10.1201/9781003088776-14

behind a semi-circular, curved table. A well-known centre for culture and politics hosts the event, including online streaming. In front of the panellists are two pedestals: one with a white *HiperSpeaker* and one with a black *HypeSeeker*.

'We are particularly interested in technologies that we choose to put in our homes. Or indeed, choose *not* to put in our homes. Per, what did you bring for us?'

'This is a *HiperSpeaker*. Since it's been on the market, less than a year ago, many people bought one. I have one at home, obviously, and I am genuinely enjoying it. Whatever my mood, at any given time, it always magically comes up with music that fits my mood'.

Mikael tilts his head and pretends to be surprised. 'Ah, magic, I see. Please do tell us about the sensors and the algorithm'.

'It has an array of cameras. Also infrared, so they also work in the dark. They sense temperatures on various body zones. They also sense your heartrate, blood saturation, eye movements and blinking patterns, facial expressions. And there is this huge catalogue of music, which the algorithm picks from. The more you use it, the better its matches. Also, the devices tap into a cloud service, so that any user can benefit from the matches that were made for other users'.

'And privacy is taken care of, I guess?' Mikael looks at Sofia.

'Sure, of course'. Per is quick to reply. 'Our systems comply to the newest regulations'.

'And Sofia, would you please tell us about your project?'

'So, this is the *HypeSeeker*. I used the *HiperSpeaker* hardware and wrote new software. It has the same sensors, but instead of playing music, it *speaks* to you. Instead of dulling your emotions with canned music, it invites you to become more consciously aware of your emotions. So you can deal with them in more healthy ways'.

Annika rolls her eyes rather demonstratively. 'That sounds even worse. A machine that *pretends* to engage in meaningful conversation. I'd rather prefer the magic jukebox'.

Mikael looks at Sofia and smiles. He's happy with some sparks in the conversation. 'Sofia, I know that you are concerned about *attention*. The ways in which, very often, technology kills our ability to pay attention, to ourselves, to others. Your prototype does the opposite. It draws attention to human-to-human relationships, to person-to-person interaction'.

Annika rolls her eyes again, this time figuratively. 'Maybe she can explain why her *HypeSeeker* is not worse than the *HiperSpeaker*? I mean, the *HiperSpeaker* violates a host of privacy rights, regardless of all the legal wizardry of their company. They make enough money to pay really good lawyers. This surveillance gadget eradicates the entire notion of privacy. It measures your emotional states 24/7 and sells your most vulnerable moments to whatever company, that can then sell you whatever product or service you're likely to buy at that moment'.

Per tries to say something but Annika continues. She turns to Sofia. 'And your *HypeSeeker* takes the dystopia even further. It targets human-to-human communication. See how far we've come? Having to qualify communication as *human-to-human*. As if machines are able to engage in genuine communication. This *HypeSeeker* not only monitors your emotions, but your private conversations, and extracts additional data, to monetize and profit from'.

Sofia remains perfectly calm; she is used to pitching her work. 'It is a prototype. Not a product. It is a *conversation piece*, meant to spark precisely the type of conversation we're now having. I would have thought that would have been clear'.

Mikael looks at Sofia and tries to hide his amusement. Or is it admiration? He turns to Per. 'You did want to say something? What *is* your company's business model?' 'Well, what I wanted to say. You can switch off the sensors and use it to play your own play list'.

'A privilege for which you will need to pay 5 euro per month', Sofia interrupts.

'Which is a reasonable price point'.

'Which is evidence of how little monetary value you assign to total intrusion in people's homes and private lives. Wow, only 5 euro for 24/7 surveillance'.

Mikael is still looking at Sofia. 'Let's look again at your prototype. If I feel, say, angry, it would detect this, and it would prompt me to do some breathing exercise. Is that correct?'

'That is correct. It would help you to first direct your attention to your own body and feelings. And then to be open to seek interactions with your surroundings, with others, with nature. We need to free ourselves of big tech's claws and reconnect to our humanity'.

Per says nothing, rather wisely, probably. Instead, he fumbles with the microphone that is clipped to his shirt.

Sofia flings her hair. 'I believe it can function as a conviviality machine. If I'm aware of *my* emotions and you are aware of *your* emotions, we can have a genuine encounter, so we can deepen our relationship'. She turns to Mikael and notices him blushing, ever so lightly.

Chapter 13

Relationships and care

The two ethical traditions that we discussed so far, consequentialism and deontology, were developed during the European Enlightenment, although their roots can go back millennia. Typical for that period were assumptions and ideals about objectivity, rationality, independence, and universality. Consequentialism assumes that you can objectively assess the pleasures and pains that follow from an action and that you can rationally add and subtract pleasures and pains, like you can add and subtract meters or kilograms. Deontology assumes that you are an independent individual and that you ought to apply rationality and follow universal moral laws. There are, however, alternatives to these two perspectives.

DOI: 10.1201/9781003088776-15

Some alternatives were developed outside the context of the European Enlightenment and, therefore, can start with very different assumptions. Here, we can think of wisdom traditions of other continents and of indigenous cultures. There is much to learn from such traditions. Other alternatives directly challenge Enlightenment assumptions and ideals. Here, we can think of perspectives that critique and propose alternatives to methods and processes that prioritize efficiency, bureaucracy, and all sorts of mechanistic or managerial approaches, notably in domains that do require human-to-human interaction, like education or health care.

I will present and discuss some of these alternatives under the broad header of *relational ethics*.

My aim will be to scratch the surface of diverse traditions and perspectives, fully aware of my inability to do justice to the depth and nuance of these very rich traditions and perspectives. Possibly, in my presentation, they come across as a motley crew. There is, however, a recurring theme: the value that these traditions and perspectives place on relationships: between people and between people and non-human animals, plants, and the natural environment.

In relational ethics, we understand people as relational beings; ethics happens within and in-between people. It values the relationships between people and between people and their natural environment, and it focuses on topics like care, communication, and collaboration.

Many traditions, across centuries and across continents, have understood the human condition as fundamentally *relational*. Below, I will review some of these, in the form of a tasting, with tiny bites from different continents.

Indigenous knowledge

Several cultures in sub-Saharan Africa, for example, follow some version of Ubuntu, a philosophy that recognizes the humanity of a person through that person's relationships to other persons. This philosophy has been a key tenet in the work of the *Truth and Reconciliation Commission* in South Africa in the mid-1990s and in Bishop Desmond Tutu's leadership in this commission. When US President Barack Obama spoke at Nelson Mandela's memorial, he reminded us that Ubuntu recognizes 'that there is a oneness to humanity; that we achieve ourselves by sharing ourselves with others, and caring for those around us'. Some researchers have applied the philosophy of Ubuntu to the design and application of technologies, notably of AI.

Many aboriginal cultures understand the human condition as grounded in connections to others and to nature and value relationships, notably connections of kinship. Australian scholar and artist Tyson Yunkaporta writes about this in *Sand Talk*. He presents an aboriginal worldview to look at global issues, sustainability in particular. He writes about how we can create, share, and store knowledge through action, interaction, and embodiment, for example, by talking while walking through the landscape, or by engaging in a *yarn*, a gathering and process of collective sense making. The book invites readers to engage with embodied learning. For example, Yunkaporta, summarizes five different ways of knowing, by inviting readers to connect these to the fingers of their hand, as tactile reminders of practical wisdom. Your little finger represents a child, to remind you of kinship-mind. Your ring finger a mother, for story-mind. Your middle finger a man, for dreaming-mind. Your index finger as the man's brother's child, to remind you of ancestor-mind. And your thumb, opposable, so that it can touch all the other fingers, can remind you of pattern-mind, to see the whole and not just the parts. (Reading this summary is, by the way, very different from engaging with the powerful storytelling of Yunkaporta.)

Other examples of indigenous traditions that value our connections to nature are found in North America. In her book *Braiding Sweetgrass*, Robin Wall Kimmerer talks, for example, about the 'honorable harvest': take only what you need; never take more than half; leave some for others; harvest in a way that minimizes harm; use it respectfully; never waste what you have taken; share; and give thanks for what you have been given. Many of the stories she relays are, on the surface level, about plants, but they pertain to much more than plants. One story is about corn, bean, and pumpkin. Indigenous peoples grow these vegetables in a mixed-crop fashion. Which is more wholesome than growing them as mono-crops. The corn starts to grow first and fast, and makes a long and firm stem. The bean can then grow and use this stem as a scaffold to climb. The pumpkin germinates last and grows large leaves, just above the ground. These leaves cover the soil and protect all three plants from draught. The plants support each other as if they are 'three sisters'. This is only one of the many examples of the connections and interdependencies between different species and between people and nature, that indigenous wisdom speaks about.

In addition, indigenous peoples in South America follow some form of *buen vivir* (or *sumac kawsay*, a neologism in Quechua, an indigenous language family in the Peruvian Andes). This translates into *good living* or *plentiful living*. This outlook recognizes the importance of our connections to nature (Mother Nature, *Pacha Mama*) and of living well, together with others. It aims to bring into balance feeling well, thinking well, and doing well. Ideas like this have informed political activism across the globe and have led to the *Universal Declaration of the Rights of Nature* by the *International Union for Conservation of Nature* in 2012 and to the inclusion of rights of indigenous peoples and of elements of the natural environment in national legislations of several countries, for example, in Ecuador and Bolivia. One form this can take is that a group of indigenous people acts as a legal custodian for one specific river or lake.

Many cultures in Asia also focus on relationships and on collective concerns, rather than on individual concerns. The Confucian tradition in China, for example, stresses relationships between people, notably those between ruler and subject, father and son, husband and wife, elder brother and younger brother, and between friends. It is possible to critique assumptions about the distribution of power in these relationships, but that is currently *not* my focus. I want to draw attention to various ethical views that focus on relationships and on collective concerns, in order to critique or complement views that focus on people as independent individuals. Because the latter have serious downsides and risks, which we will discuss in the next section.

Many associate relational ethics with an orientation towards others and with care for others. In addition, I would like to propose that it also involves an orientation towards oneself, to acknowledge that one's being human is grounded in relatedness. To embrace the human condition of being a relational being, to treat oneself with care, can promote wellbeing, both of others and of oneself. Conversely, treating oneself as an isolated, unrelated individual can corrode one's humanity.

European Enlightenment

In this chapter's introduction, I remarked that relational ethics can be understood as challenging key assumptions and ideals of the European Enlightenment. Now, what are these?

A key feature of the Enlightenment is the development and indeed combination of science and technology, of politics and economics. If we are looking for a starting point for this development, we can point at British philosopher and statesman Francis Bacon (1561–1626). Bacon played a key role in the creation of the Royal Society, a major engine of the development of modern science and technology. Bacon promoted a view of man separate from nature and an instrumental approach towards nature; as a resource that can be exploited. 'Nature must be taken by the forelock', he notoriously wrote, invoking an image of a man assaulting nature. 'Knowledge is power', he also proposed. If we subjugate nature, we can glean knowledge from her and use her for our own purposes, notably to increase power. This approach has been further developed since. It is the taken-for-granted foundation of untethered neo-liberal politics and economics, which dictates the exploitation of nature and of people, in grimy sweatshops and in exhausting gig work, and the rampant extraction of value from nature.

In our current time, many people have lost contact with nature. Do you know where the food came from that you find in your shop? Where did the ingredients come from? Likewise, many people have lost contact with what they believe to be 'others', anyone 'different' from them. Modern science, technology, and neo-liberalist politics and economics have been combined into a deadly mix and have led to the climate crisis, economic inequalities, and global injustices.

The separations between people and nature, and between people, are key themes in the work of Otto Scharmer, senior lecturer at the Massachusetts Institute of Technology (MIT). Scharmer has developed *Theory U*, an approach to organizational change and innovation. He argues that we have, largely unintentionally, created three divides, which form the root causes for our global problems. An *ecological divide*, a way of looking at the world and dealing with nature, as if we are separate from it and can exploit it at will. A *social divide*, a way of looking at other people in terms of 'us' and 'them' and a belief that huge inequalities between groups are acceptable. And a *spiritual-cultural divide*, a disconnect between what you currently think, feel, and do, and another potential version of you, where you have cultivated virtues like courage, self-control, justice, and wisdom. In our work, we can heal and close these divides, and create a *fertile soil* (a word that Scharmer uses; he grew up on a farm) for collaboration towards positive and systemic change. Likewise, my hope is that we can learn to connect to nature, to others, and to our better selves, and experience that we are part of this incredibly awesome web of nature, part of this hugely diverse human family, and can steer technology and innovation in beneficial directions. There are plentiful initiatives for ecological and social justice and reform. You can think of examples that are relevant to you and that you may want to support or contribute to.

Contemporary thinkers

Views on the human condition that put relationships central can also be found in contemporary philosophy. Emmanuel Levinas (1906–1995), for example, wrote about the encounter between *Self* and *Other* and about the responsibility that follows from

this. It can also be found in *communitarianism*, which empha-
sizes the role of community and advocates creating and protecting institutes that foster relationships. One can also think
of Michael Sandel (born 1953), professor at Harvard University,
known from his popular online course and book *Justice: What's
the Right Thing to Do*, or of Alasdair MacIntyre (born 1929),
who wrote *After Virtue* (1981), which contributed to the rekin-
dling of virtue ethics (see Chapter 15).

Furthermore, there are insights from biology and from the
social sciences that support a view of people as inherently
relational. Primatologist Frans de Waal, for example, wrote
extensively about the tendencies in our primate-cousins to collaborate, rather than compete. For us primates, 'survival of the
fittest' refers to our capabilities to collaborate in and between
groups and to survive as groups; it does *not* refer to competition between isolated individuals. Similarly, historian Rutger
Bregman, in his book *Humankind: A Hopeful History*, presents
evidence that most people, most of the time, do want to collaborate and be kind. He debunks a series of infamous psychology experiments, like the *Stanford prison experiment*, in which
people were incorrectly presented as prone to cruelty. One may
wonder where the belief that people are cruel comes from. One
explanation is that many news outlets, and especially so-called
'social' media, have incentives to bring stories of crime, vio-
lence, and terrorism. These events, however, are the exceptions,
not the rule. Many outlets and media make money by selling
ads, and ads sell better to people who are 'engaged'. And dis-
content, outrage, and hostility fuel 'engagement'. This type of
(fake) news distorts people's outlook on the human condition.

Ethics of care

Moreover, in the last quarter of the 20th century, a strand in
normative ethics has been developed that takes relationships
as a starting point. Carol Gilligan (born 1936), Nel Noddings
(born 1929), Judith Butler (born 1956), and others have developed *feminist ethics* or *ethics of care*. They view people as
mutually dependent. The dependency is most obvious in the

relationships between parents and children and between people who give care and people who receive care.

But relationships are also at play between family members, friends, co-workers, or business partners. This relational view acknowledges that we are not only rational beings, but also emotional beings. When you reflect on a situation and deliberate what to do, you do not only need to use your faculties for thinking; you also need to use your faculties for feeling, like empathy or compassion. Moreover, we need to take into account that each situation is embedded in a specific context and take into account these specifics. This offers an alternative to the Enlightenment's assumptions and ideals of objectivity, rationality, independence, and universality.

In order to compare and contrast care ethics with other ethical traditions, we can look at how it deals with the concept of *care*, in relation to *justice*, a key concept in all ethical traditions. In consequentialism, justice entails distributing good and bad outcomes fairly among different people and groups. In deontology, justice is understood as protecting human dignity, autonomy, and rights. Now, supporters of ethics of care would argue that both care and justice are needed. In the judiciary system (an example of a 'public' context), judges are primarily concerned with justice. It would, however, be wrong if they did not do that with care. They need to look carefully at the people who appear before them and at, for example, the specifics of their personal circumstances, motivations, and actions. Conversely, in another context, say, the context of living together in a family (a 'private' context), it is very likely that care is a leading principle. It would, however, be wrong if family members ignore basic concerns for justice, for example, when they share goods. Fairness and equality are necessary in family life. In ethics of care, both justice and care are needed and function as complementary concerns.

When we want to distribute good and bad consequences fairly between people, we need to look carefully at who are at the receiving end. If they need care, you need to take their needs into account. Likewise, respect for dignity, autonomy, and rights need to go hand in hand with taking into account the specific circumstances of the people involved. These are natural and cultural impulses. This is how relationships, families, groups, and societies work. Sometimes *equality* is required, in the sense of treating different people equally. Sometimes *equity* is required, in the sense of treating different people differently, depending on their needs.

Ethics of care emerged in 'personal' contexts, in health care relationships or in small groups, like families, but has since been developed to speak also to 'public' contexts, like economics, politics, law, ecology, and international affairs. The words 'personal' and 'public' are between inverted commas because we can question whether these domains are totally separate.

If we take ethics of care as a starting point, societies would need to be organized very differently. Many societies currently value employment in companies and earning money, which is often done by men, more than work in informal settings and providing care, which is often done by women. In her book *Ethics of Care*, Petra Held envisions a different society, in which 'bringing up children and fostering trust between members of society [are] the most important concerns of all. Other arrangements might then be evaluated in terms of how well or badly they contribute to the flourishing of children and the health of social relations. That would certainly require the radical restructuring of society! Just imagine reversing the salaries of business executives and those of child care workers'. Similar notions came to the fore during the Covid-19 pandemic, in discussions about work that people do in health care, education, cleaning, logistics, and retail; essential work that too often receives too little appreciation or compensation.

Relations to other perspectives

The role of relational ethics or ethics of care is debated. Some propose that it is a separate perspective. Others suggest that it can be used in combination with other perspectives; for example, to enrich these other perspectives. Below, I will explore the latter approach.

We already discussed the need to assess potential positive and negative consequences (Chapter 9). We can add to that the need to assess potential consequences for the qualities of various relationships. When people use the system that you work on, how could that affect the quality of the relationships or interactions this system facilitates? Does it, for example, promote viewing others as persons, in face-to-face encounters? Or does it reduce people to data points and corrode human-to-human connections and interactions?

Likewise, we can add to our discussion of duties and rights (Chapter 11), the duty to take care for fellow human beings, including those on other continents and of future generations, for non-human animals, and for the natural environment on which our lives depend. They are all 'others', removed in space or time, who have moral or legal rights. The duty to take care is paramount in the climate crisis in terms of distributive and intergenerational justice.

German-American philosopher Hans Jonas (1903–1993) wrote about some of these topics in *The imperative of responsibility* (1984). He formulated the following imperative: 'Act so that the effects of your action are compatible with the permanence of genuine human life'. This is a combination of duty and care. We can find this imperative also in the requirement for sustainable development, which was defined as follows, in *The Brundtland Report* (1987): 'Development that meets the needs of the present without compromising the ability of future generations to meet their own needs'.

Smart devices in our homes

Let us look at technology that intrudes in our daily lives, that influences the ways in which we interact with each other. What happens when people put so-called 'smart' devices in their homes? I refer here to devices that are connected to the internet and that are equipped with sensors, like a microphone or camera, with processing capacity, for example, to recognize speech, and with actuators, to perform simple acts. A speaker, equipped with voice recognition that answers questions by looking-up information online. A doorbell that enables you to see who is at the door while you are out and to open the door to let them in. Or the refrigerator that keeps track of stock and orders goods online; an appliance that has appeared since the 1990s in then-futuristic Internet-of-Things scenarios.

How do such devices affect relationships? Can they help to improve relationships? Or do they corrode them, for example, because people continuously stare at their devices' screens instead of look at the other's face? These devices, and the software and services on them, are typically programmed to grab and hold our attention. And using them does affect our abilities to concentrate, to engage in reflection, and to communicate with others. Journalist Nicolas Carr, as early as 2010, wrote about this in *The Shallows: What the Internet is doing to our brains.*

Of course, it depends on how people actually use these devices. Here is an interesting role for designers and developers. They make decisions regarding the features of such devices. And these features influence how people use these devices. Imagine, for example, that you include a feature that enables people be more aware of the time they spend with their device. They can set a timer for a number of minutes. When the timer goes off, it reminds them of their intention to use their time wisely. Such features enable people to cultivate self-control (see Chapter 15). Too often, however, these 'smart' devices are designed to hijack people's abilities to exercise self-control. These so-called 'tools' then turn into tools that companies can use to grab our attention and manipulate our actions.

Practical applications

How would you go about if you want to apply relational ethics in your project? At first glance, it might seem rather unhelpful. It offers no plusses or minuses. No universal duties or rights. Relational ethics purposely stays away from objectification and generalization. Relational ethics can, however, help to focus on the impacts of your project's outcomes on relationships between people and between people and the natural environment.

For many years, I worked on projects in the telecommunication industry, which is all about interactions between people.

Thus, the focus of relational ethics can coincide with the objectives of projects in this industry. In one project (TA2), we created prototypes that were meant to connect family members or friends, to promote experiences of togetherness. We organized an iterative process, in which we developed a series of demonstrators and prototypes, which we used to study different users' experiences. It was great to see that the developers and engineers were very interested in the interviews, workshops, and experiments that we did with potential users.

In another project (WeCare), we developed several online services to support people who provide informal care to people who suffer from dementia. Often, their parents. Needless to say that this is a sensitive domain. Moreover, the work of providing informal care can be stressful. Consequently, many informal carers suffer from burn-out. Our aim was to create applications to support them, for example, in coordinating care tasks between each other and with professional care workers. This project was not about automating tasks, but about supporting people and enabling them to interact. Without knowing, we did something like relational ethics.

Relational ethics also invites you to envision how technologies, and their effects on relationships and interactions, can affect larger societal, economic, cultural, or political structures and institutions.

Interestingly, scholars like Abeba Birhane, Mark Coeckelbergh, Virginia Dignum, Charles Ess, and David Gunkel have proposed relational approaches to the design and application of robots and other autonomous systems. Coeckelbergh, for example, proposed to adopt three perspectives: a *first-person* perspective, of the person who interacts with the non-human entity; a *second-person* perspective, of the non-human entity; and a perspective that looks at the *interactions*, the shaping of habits, patterns, and relationships, as they develop over time. Such a relational view offers an alternative to viewing technologies as passive tools that people can pick-up and use or as autonomous agents, which we need to monitor and control.

Rather, a *relational* approach focuses on the ways in which people and technologies interact and shape reciprocal relationships between them. In a similar vein, Kate Darling, a researcher of legal and ethical implications of technology at the MIT Media Lab, in her book *The new breed*, proposed to learn from the ways people have, for millennia, interacted with, for example, dogs and horses, and to use insights from these practices in how we design interactions with robots.

Let's make this exercise a bit more challenging. Let's introduce several layers of relationships. Take a project that you work on. Now, how do you interact with the other people involved? What do you think and feel about these interactions? Anything that may be improved regarding communication, collaboration, or care? You can enlarge your circle of moral concern by envisioning your project's outcomes and impacts on the world. Which people are affected by it? Which groups of people can you think of? And how may the product, service, or system that you work on affect the ways in which these people or groups interact with each other? How would it affect people's abilities to communicate or collaborate with others, or to provide and receive care? How can the deployment or usage of the system that you work on affect the ways in which people relate to other people, and how can it affect the ways in which people relate to nature? Bonus exercise: Explore how you can combine concerns for justice with concerns for care.

Chapter 14

A social media app

Stan is a social entrepreneur. His start-up develops a social media app that will enable people to share durable goods, such as power tools and other do-it-yourself equipment, and also to share experiences, to offer and to receive help, and to collaborate, for example, with neighbours.

They have an opportunity to meet an investor next month. The four of them are working on a prototype that they can pitch to him. This is his second start-up; his first crashed two years ago for lack of funding.

It's lunchtime. They went downstairs to the common room of the social innovation hub where they rent office space. There's Jelle, a senior back-end programmer with a drive to make the

DOI: 10.1201/9781003088776-16

world a better place, Maaike, a front-end developer, hardwork-ing and keen to make money, and Lisa, user experience, design, and testing, the youngest and very bright—this is her first job.

They sit around a round wooden table with various ceramic plates and bowls full of organic food: fresh vegetables and sal-ads, rice and baked sweet potatoes, fruits and nuts.

Lisa brought a print-out, which she manages to find space for in the middle of the table. 'On this slide, you see people's responses to the three design options. Clearly, the blue one was most popular. Moreover, we can use some of the feedback on the other options to further improve it'.

Stan stares out the window; he's already seen the graph. 'Great, great'. He turns to Lisa.

Maaike moves and straightens her spine. 'I've helped with getting everything to work. We can implement these changes within two weeks, well in time before the pitch'.

'Yes, well'. Stan wipes his mouth with a napkin. 'I've been thinking about this investor. Also, I talked to some good friends of mine. They asked me a couple of questions that made me think. What is that we're trying to build? What do we want to put out into the world? Why would we want to sell that to somebody else?'

Maaike puts down her knife and fork. 'Really? You want to delay? What do you mean? I mean we've got to move on, right? Build and deliver. This is our window of opportunity'.

Jelle has been collaborating with Stan for some years. 'Questions, about what? What do you worry about?' 'Okay, well'. Stan needs a bit of runway before take-off. 'Look. We've been working like crazy for three months. The three of you, I mean. And you've done an excellent job. That is not my worry. But why *would* we? For the money? Do we desperately need money?'

'Maybe not desperately'. Maaike looks around at the others. 'But I guess we would want to have decent compensation for these three months. More than decent would be even better'.

'Of course, I understand and I agree'. Stan searches for words. 'What I am thinking is: What it will do to *us*, as a team, and what it will do to our product? I mean, if we let this investor in'.

'What will happen is: lots of people will benefit from a great service. And the environment will benefit, and social cohesion will improve. That is our mission, right?' asks Maaike, rhetorically. 'And we make some money in the process. I don't get it. What *is* your problem, Stan?'

'This friend of mine, we've known each other since university, asked me about what I want my life to mean. Sure, we can work on this. But for *what*?'

Jelle nods, swallows the food and clears his throat. 'If I can speak candidly, I've never been very fond of this investor. Why not keep it small and simple? Grow organically'.

Maaike is still looking at Stan. 'What are you trying to tell us? Do you want to quit? Do you want to proceed?'

'I imagined myself in five years' time'. Stan has a puzzled look on his face. 'What type of company we can have. The service we offer. What it does for people, for the environment. And to be honest, I do have second thoughts about this investor and their influence on our work. What they want to do with our product may differ from what we envision'.

Maaike takes a spoon full of fruits and nuts and places it on her plate. 'Well, I *do* want to proceed. I want our company to be a success and we need an investor for that'.

Lisa has been silent so far. 'He isn't saying he does not want the investor. He's just thinking out loud, right?'

'In a way, yes'. Stan takes a sip of water. 'To me, there is more than plusses and minuses. I say we give it a couple of days. I do understand that we need to make a decision about the investor rather soon. But not today. For me, it can go both ways. Either it feels good to proceed with this investor. Or we feel it's not the right thing to do, at least for now'.

There is a moment of silence. Maaike taps her hands on the table top. 'Well, maybe that is not exactly democratic. But I hope that you do make up your mind this week. Meanwhile, Lisa and I will proceed working on the prototype. Right?'

'Sure. And, what I wanted to say', Lisa tilts her head, 'Maybe we can just ask the investor, his name is Adrian, right? We can ask Adrian. What he wants to do with the product. What type of company he wants to see us grow into'. Stan nods and Lisa continues. 'I know you did discuss that with him. But that was three months ago. Things may have changed. Also for him. Can we help you prepare and plan such a discussion? Maybe Maaike and I can prepare questions about stocks and revenues. And maybe Jelle wants to prepare questions about vision and roadmaps'.

Jelle moves his body in the chair, lets his back relax against the chair's backrest. 'Excellent proposal, Lisa. Stan, what do you think? Maaike, are you okay with this? Tomorrow we schedule some time to flesh-out a couple of questions we would like to discuss with Adrian. Now that I think of it, I can envision Adrian actually enjoying such a discussion'.

Chapter 15

Virtues and flourishing

Virtue ethics has ancient roots. Different versions of virtue ethics have emerged in different cultures. The Western tradition of virtue ethics originated in ancient Athens, with Aristoteles (384–322 BCE) as a key figure. Aristotle was a student of Plato and later founded his own school, the *Lyceum*, where he taught Athens' citizens what we would today call a combination of political philosophy and moral philosophy. Other virtue ethics traditions are, for example, Confucian or Buddhist, which were developed one or two centuries before Aristotle. Virtue ethics may sound antique and irrelevant for our time, but many people, in academia, in industry, and in popular media, believe it is worthwhile to turn to it and revive it, for example, in business

DOI: 10.1201/9781003088776-17

and professional ethics, and in self-help books on 'the good life' or 'the art of living'.

Some people misunderstand virtue ethics. They believe that it is individualistic and deals with living a goody-goody life. That is not the case. In order to correct such misunderstandings, let me take you back to Aristotle's Athens. Let us dig up the ideas of *polis*, *telos*, and *ethos*; ideas that have gone missing in the centuries since. Let us blow off the dirt and the dust and see how we can use these ideas in our time. (Please note that some assumptions of Aristotle's time, notably regarding slavery and women's rights, are plainly unjust, and, fortunately, have been corrected over time, at least partially.)

Polis

Aristotle's teachings were concerned with living together in a *polis*, a city. He believed that we are *zoon politikon*, social animals. We are meant to live together. That idea got lost during the Enlightenment. Many people currently believe that we are independent individuals. Losing the idea of community is unfortunate and has been a major cause of lots of injustice.

Virtue ethics does not ask how I, as an independent individual, ought to act, but asks what I can do to contribute to creating conditions in which people, including myself, can *flourish*. Of course, our time is very different from Aristotle's. He was concerned with living in a *polis* with tens or hundreds of thousands of people (many without citizenship; notoriously, women and enslaved people). In contrast, we are concerned with living together with billions of people on one planet and with global issues, like the climate crisis. Or, on the level of nation states, we are concerned with finding ways to live together with millions of people, with whom we share legislation, language, culture, and institutions. We can also be concerned with collaboration between countries. I sometimes imagine the European Union as a *polis*, with over 400 million people in 27 different countries. This image immediately raises a range of questions, like: Where are the boundaries of this *polis*? How solid or fluid should we make these boundaries? Which people can or cannot be citizens of the EU? And how to balance national and international concerns? Such questions are at play, for instance, in deliberations about migrants or refugees, and about solidarity between member states.

In short, virtue ethics is inherently social and virtues are meant to promote conviviality, to find ways to live well *together*. You may have noticed that I have used the term *flourish* several times already. It is meant to convey a key idea of Aristotle: *eudaimonia*, which refers to living well together. It includes happiness, but it is much more than 'feeling happy'. We can understand *eudaimonia* as living a meaningful and fulfilling life, together with others. We can characterize such a life as satisfactory, with pleasant and happy moments, but also with moments of difficulty and pain.

Telos

Another key concept of Aristotle is *telos* or purpose. I can illustrate what he meant with that by asking: *Why does it rain?* When you hear this question, what do you think of? Possibly an explanation; something with water in the seas and oceans, sunlight that evaporates the water, formation of clouds, a temperature drop, and the creation of raindrops. Maybe you heard the question as: *Where does rain come from?* Aristotle, however, would hear the question differently. For him, everything and everybody has a particular purpose, a *telos*, and everything and everybody strives towards their particular purpose. He would hear the question as: *What is the purpose of rain?* Well, it rains so that plants can grow and plants can feed animals and people. An acorn strives to grow into an oak tree. People strive towards realizing their human potential to flourish. And collectively, our purpose is to create societies in which all people can flourish.

This idea of purpose, however, also went missing in the Enlightenment, which was all about mastery, about means and mechanics. People made huge advances in science and technology, which yielded a focus on exercising control and on growth for its own sake; more of the same. 'We've constructed ourselves an industrial system that is brilliant on means, but pretty hopeless when it comes to ends', wrote John Thackara, a pioneer and critic of online media; 'We can deliver amazing performance, but we are increasingly at a loss to understand what to make and why'. For Aristotle, the end was obvious: human flourishing to live well together.

Ethos

A third concept of Aristotle that we need to understand is *ethos* or virtue. Some people recoil at the notion of virtue because it reminds them of being told to behave well and comply to all sorts of norms. That is *not* what Aristotle taught. Quite the opposite. In *Nicomachean Ethics*, Aristotle explained virtues in terms of *excellence* (which resonates in the word *virtuoso*). A virtue is a *disposition* that one develops based on past actions and that guides one's future actions. It is a disposition that aims for an *excellent* expression of relevant virtues, according to the *appropriate mean*. That is a mean between deficiency and excess: between doing too little or too much of something, dependent on the specific, practical situation one finds oneself in. You can develop and cultivate virtues *over time*, so that your thinking, your feeling, and your acting become more and more aligned. The purpose is to increasingly align your head, your heart, and your hands.

You may understand Aristotle's advocacy to cultivate virtues as 'fake it until you make it'. He advocated acting virtuously, embodying a virtue, as a step towards cultivating virtues. It is easier to *act* your way into a new way of thinking than to *think* your way into a new way of acting.

Let us turn to an example: the virtue of courage; the excellent expression of the appropriate mean between cowardice (deficiency) and recklessness (excess). Imagine that you are out on the street at night. You see a person being attacked. If you are a frail person, it would be courageous to stay out of the conflict and phone the emergency number. It would be reckless to intervene in the fight. Alternatively, if you are an athletic person and skilled in deescalating conflict, intervening would be courageous. It would be cowardly to stay out of the conflict.

So, different people in different situations will need to find out, and try-out, what would be an appropriate expression of a particular virtue, depending on their particular abilities and on the specific situation.

Notably, one particular virtue can take very different shapes in different contexts. Courage, for example, looks different in firefighting, in nursing, and in data science. A firefighter walks into a burning building to rescue people, fully aware of the danger, having trained extensively to work in such circumstances. A nurse often may need courage in the form of endurance in order to provide care in adverse circumstances. A data scientist may need to ask uneasy questions about difficult issues and tolerate awkward situations resulting from this. What these different expressions of the same virtue share is acting courageously in the face of danger, adversity, or difficulty.

It is important to stress that this *mean* has nothing to do with mediocrity. The mean for courage is *not* between too little courage and too much courage. Rather, it aims at excellence: doing very well what a virtuous person would do in this specific situation. One can find this mean by using practical wisdom (*phronesis*). What would a courageous person *do* in this situation? Or a *just* person? Or a *compassionate* person?

Aristotle gave practical advice to find this mean. He advised those who tend to go to the excess-side of a specific virtue, to try and go to what feels like the deficiency-side of that virtue, in order to hit the appropriate mean. If you found yourself acting rashly before, do something that may feel like cowardice (to hit the mean). Alternatively, if you tend to go to the deficiency-side, you may want to try out what feels like the excess-side. If you found yourself acting cowardly before, do something that may feel like rashness (to hit the mean). These examples illustrate that self-awareness, self-knowledge, and reflexivity are critical ingredients of practical wisdom. You need wisdom to steer and modulate other virtues, like courage, self-control, or justice.

Finally, it is important to understand that virtue ethics is everything but restrictive. The opposite is the case. It is aspirational. It does not state what you should or should not do. Rather, virtue ethics helps one to guide and shape one's natural impulses; to shape and follow one's desires, like the desire to eat well and take care of your body, the desire to create beauty, or to contribute to a larger purpose. In that sense, virtue ethics is an alternative to consequentialism, which can involve calculation and making trade-offs, and to deontology, which can involve compulsion and curtailing oneself.

Furthermore, in its acknowledgement of the specifics and complexity of each situation, virtue ethics is especially useful for professionals, who find themselves in specific and complex situations, often with intricate histories and conflicting demands. Moreover, there is a time-dimension to virtue ethics; cultivating virtues takes time and involves learning from experiences.

In virtue ethics, we look at people as rational, experiential, social, and active beings. It focuses on cultivating virtues (ethos), finding the appropriate mean in specific situations, on growth (telos), and on living well together (polis) and promoting people's flourishing (eudaimonia).

Practical applications

Now, how would you apply virtue ethics in your projects? Virtue ethics is *not* about counting plusses and minuses. It is also *not* about clear rules that apply universally. In virtue ethics, it depends. It depends on the specifics of the situation and on the people involved and their abilities and willingness to cultivate and exercise relevant virtues. It depends on the various conceptions that people can have of the good life, of the kind of society they want to create.

Say, you are involved in the design or application of an algorithm for a government agency. The algorithm is meant to find fraudulent behaviour of citizens. Which virtues would you need in such a project? Self-control? Courage? Justice? Humility? You can pick one or two of these virtues and create opportunities to cultivate them. You may want to say 'no' to a proposal to add some functionality that stretches the project's scope towards potential misuse. This would involve self-control and courage. Self-control in that it restricts what will be built. And courage in that it can be difficult to argue against a prevalent logic of adding functionalities. Additionally, you may want to carry out an experiment to evaluate the algorithm's fairness. You may want to look into not only the fairness of the algorithm in a narrow sense, but also into the fairness of the processes around the algorithm, for example, whether operators or citizens are able to inspect the algorithm's functioning or to correct the algorithm's output. This would involve not only justice, but also humility, for example, in being transparent about what the algorithm can and cannot do, acknowledging its limitations.

It is critical to experiment and to learn from your experiences, when you want to cultivate virtues. At first, you may feel awkward or your actions may not be entirely successful. Over time, however, you will become better in it, and your thoughts and feelings will better align with your acts. And the other way around. You may need to consciously modify your behaviour and you will probably still feel somewhat uneasy, but you will act more virtuously nevertheless.

Cultivating a virtue is a process. A person who has cultivated a virtue will have learned to express this or that virtue out of habit, in an optimal form, for appropriate reasons, and with appropriate feelings. This is the beauty of virtue ethics: it involves an aspirational mindset; it enables exercise, learning, and growth.

In addition, you can look at *exemplars*, people who embody, exemplify, or champion specific virtues; people whom you can look to as role models. In Chapter 21, you can find a series of short profiles of people whom I see as *exemplars*. You can watch online talks by them or listen to interviews with them. You can model their ways of thinking, feeling, and acting and learn from them. You can try-out relevant virtues in your work and learn by doing.

Contemporary virtues

Aristotle discussed a range of virtues that he found relevant for citizens of Athens. For each virtue, he discussed the well-cultivated, excellent form, and less appropriate, deficient or excessive, forms. Many of these virtues are still relevant today, like courage, self-control, justice, and wisdom. Each virtue refers to a one's disposition to reliably find a mean between excess and deficiency that is appropriate, given the context and one's abilities.

In her book *Technology and the virtues: A philosophical guide to a future worth wanting*, Shannon Vallor, Baillie Gifford Chair in the Ethics of Data and Artificial Intelligence at the Edinburgh Futures Institute at the University of Edinburgh, proposes that virtue ethics is needed to discuss the opportunities and risks of *emerging* technologies. She argues that consequentialism and deontology are less suitable because they require more clarity and more details of the technologies we deal with. *Emerging* technologies, however, are in the process of being developed so that we can know very little about how they will function practically; and even less about how people will practically use them in their daily lives. A consequentialist will, therefore, find it hard to determine and analyse the plusses and minuses of, for example, a robot system that a social enterprise is developing in order to provide care for elderly people and to support care workers. There is a wide and diverse range of actors and stakeholders involved, with various and conflicting interests. Many question marks. Likewise, a deontologist will find it hard to articulate a categorical imperative that is appropriate for, for example, a surveillance and security system that a private-public partnership is developing and that includes cameras in public spaces. Identifying and discussing all relevant duties and rights will be very challenging. Again, many question marks.

For products or services that are under development, virtue ethics offers a lens to alternately zoom-out and zoom-in. You can zoom-out to explore more general questions about the ways in which the innovation that you work on may help or hinder to create a 'future worth wanting', a society in which people can live well together and flourish. Alternatively, you can zoom-in to explore more detailed questions about specific features of the product or service you work on and explore how these may help or hinder people to cultivate relevant virtues.

A social media app

We can have a look at social media, using an example from Vallor's book. In 2013, *Facebook* aired a series of video ads to promote its *Facebook Home* service. One of the ads portrays a group of eight people, a family with an additional aunt and uncle, at a large table full of food, four on each side. The aunt tells in boring detail about what happened to her today. Some pretend to listen. Others concentrate on their food. One rolls their eyes. The video's protagonist is a teenage girl. She takes action; presumably the action that Facebook endorses. She pulls out her mobile phone, keeps it in her lap, and browses through a series of snapshots of her friends: a friend behind his drumkit, drumming; her friends outside, throwing snowballs. As she scrolls these photos, these people appear, magically superimposed, with sound, in the room where they have dinner. The drummer in the corner. Her friends throwing snowballs across the room. She can experience a virtual reality that is more attractive than the reality she currently is in. This, by the way, Vallor points out, only works when one person retreats to their personal bubble. If all the people at the table would retreat in their isolated bubbles, the joint dinner would collapse.

Vallor argues that the ways in which people use technologies can either help or hinder them to cultivate specific virtues. For people who work in tech, this means that they also need to cultivate a set of virtues, in order to design and apply technologies in ways that foster prospective users' abilities to cultivate specific virtues, and thus promote conditions in which people can flourish and live together well.

For the design and usage of social media, virtues like self-control, empathy, and civility are especially relevant. Vallor defines self-control as *'an exemplary ability in technomoral contexts to choose, and ideally to desire for their own sakes, those goods and experiences that most contribute to contemporary and future human flourishing'*. Social media, however, are often designed to lure you into using the app and to keep you glued to it. This easily undermines people's abilities to exercise self-control. Alternatively, you could envision and create an app that enables people to cultivate self-control, for example, to articulate their goals and work towards these goals. The interface could prompt the user: 'You have set a 5 minute limit. These 5 minutes are over. You may want to focus on your goals'.

In addition, we need empathy. The absurdity of 'social' media is that in theory they can bring people together; in practice, however, they often corrode our abilities to communicate. Vallor defined empathy as a *cultivated openness to being morally moved to caring action by the emotions of other members of our technosocial world*. Imagine that your job is to create a social media app to cultivate empathy. Such an app could display information that helps the user to better understand other people and their experiences and to connect to people with diverse experiences. It could function as a tool to become better at listening, communicating, and relating to others.

Moreover, in the context of social media, we need to cultivate the virtue of civility. Maybe you associate this term with courtesy or politeness. Vallor, however, defines it differently: as '*a sincere disposition to live well with one's fellow citizens ...; to collectively and wisely deliberate about [what matters]; to communicate, entertain, and defend our distinct conceptions of the good life; and to work cooperatively toward those goods of technosocial life that we seek and expect to share with others*'. Our social media would look very different, if they were designed and used as tools to cultivate civility.

You can think of other virtues that are at play (or at risk) in this dinner table case. For example, courage, to speak up, to express one's feelings, to go against the pattern of rolling eyes. Or care, to take action to foster more compassionate ways of interacting between the people present.

Relevant virtues

If, by chance, you find the vocabulary of *virtues* archaic, you can think of them as *superpowers*: the superpower to choose goals that contribute to flourishing (self-control); the superpower to be open to others and to act to improve their lot (empathy); or the superpower to engage in joint deliberation and collective action to find ways to live well together (civility).

There are many lists of relevant virtues. Here, I would like to mention a series of virtues that I would see as especially relevant for people who work in tech. My goal is to provide good enough starting points for cultivating these virtues, without claiming completeness or rigour.

People who develop new technologies need to cultivate (some of) these virtues, in order to deliver technologies that can support others ('users') to exercise the very same virtues. If you are working on an algorithm that can impact people's lives in terms of justice, for example, regarding fairness, and equality, then you will need to cultivate the virtue of justice. Similarly for the other virtues.

First, there are the four *cardinal* virtues: courage, self-control, justice, and wisdom. These date back to ancient Greece, but they are still relevant today:

Courage: The ability to act rightly in the face of adversity; a disposition to perceive dangers and opportunities and navigate between these; to find an appropriate balance between fears and hopes, between cowardice and rashness. Courage may include perseverance, dedication, and commitment. Courage plays both during design, for example, in mentioning some uneasy topic, and during usage, for example, in supporting people to cope with difficulties.

Self-control: The ability to steer one's desires and impulses; a disposition to choose habits and experiences that promote human flourishing. This may include temperance, discipline, and patience. As a designer or developer, you may need to exclude specific features (to combat 'feature creep'). While working on systems or products, you may need to consider how these can enable (or dampen) prospective users' abilities to exercise self-control.

Justice: The ability to notice and evaluate diverse benefits and drawbacks of specific innovations or applications; and to seek just and fair distributions of these benefits and drawbacks across people and across groups (distributive justice). You will need to consider how these innovations or applications can promote (or corrode) (material and procedural) justice, fundamental rights or human rights, and wellbeing of specific individuals or groups of people.

Practical wisdom: The ability to determine, for each specific situation, which virtues are needed, and to express these virtues appropriately, aiming for an appropriate *mean*. It functions as a *master virtue* that you can use to steer and modulate other virtues. It involves reflexivity, self-awareness, and self-knowledge, in order to critically reflect on practices in which you are involved and on your role and participation in these practices.

In her book, Vallor identifies and discusses a series of *techno-moral* virtues, which we need to cultivate in order to flourish in our current, *technosocial* world:

Honesty, reliability or integrity: A disposition to respect and promote truth and to build and promote trust. This is relevant, for example, regarding fake news and political and cultural polarization. In your projects, you may need to consider how the system or product that you work on can foster (or stunt) honesty in people who use this system or product.

Humility: A disposition to recognize the limits of science and technology. Humility would involve questioning what technology can and cannot do and would avoid an overreliance on *technological* innovation. It would, for example, help to bring attention to the potential of *social* innovation, as a complement or alternative to technological innovation.

Civility: A disposition to seek ways to live well together with others; to promote joint deliberation and collective action towards societal goods. This is different from 'being polite'. In your work, you may want to include concerns for societal goods and consider how the product that you work on can encourage (or hamper) people who use it to cultivate civility.

Empathy and compassion: A disposition to be concerned with others and with non-human animals, to be moved, and to take action. This may involve questioning your project's objective. You may want to reflect on the ways in which the product that you work on can aid (or stifle) people's abilities to exercise empathy and compassion.

Care: A disposition to meet the needs of others; to contribute to the ameliorating of suffering. This can refer both to your role in a project and to the project's outcomes. For example, you may need to consider how the product that you work on can help (or hinder) people to care for themselves, for others, and for nature.

Vallor also discusses several *technomoral* virtues, which I chose to modify, in order to adapt them to the experiences of people who work in innovation and technology projects:

Perspective and curiosity: A disposition to look at situations and appreciate the various moral elements in them; and a disposition to be open and receptive towards other people and their experiences and learning from them. This virtue is needed in organizing all sorts of meetings, both with project team members and with potential users.

Flexibility and creativity: A disposition to steer and modulate one's actions, beliefs, and feelings to changing situations; and a disposition to generate ideas and combine different ideas. I like to think of curiosity and creativity as mirror images, as complementary moves: curiosity has to do with impression (going in), creativity with expression (going out).

Lastly, I would like to mention several virtues that are relevant to working in technology and innovation projects and to engaging with prospective users or other stakeholders. We will further discuss these virtues in the chapters on Human-Centred Design (Chapter 17), Value Sensitive Design (Chapter 18), and Responsible Innovation (Chapter 19):

Collaboration: A disposition to promote and foster cooperation. This virtue is probably best expressed in combination with other virtues, for example, in *collaborative curiosity* or *collaborative creativity*, and it requires care for the people involved and for group dynamics. Collaboration plays both in the context of design and in the context of usage.

Empowerment: A disposition to view the systems that you help to develop as tools to empower their users. Empowerment requires that people in design roles enable putative 'users' to participate actively in the design process. It also involves empowering 'users' to exercise virtues like self-control, empathy, and civility, when using these systems.

Anticipation and responsiveness: A disposition to explore both desirable and undesirable outcomes of your project, including, for example, higher-order effects, and to respond to changes and findings during the innovation process, for example, to question earlier choices. It involves the organization of iterative processes of experimenting and learning.

Diversity, inclusion, and participation: A disposition to promote diversity, for example, in a project teams' composition, to include unusual perspectives or fields expertise or to enable diverse 'users' to participate in research and development, and in application and deployment. The latter requires sharing power with them (see *Empowerment*).

Take a project that you work on. Close your eyes. Inhale. Exhale slowly. Feel your feet on the ground. If you are siting, feel the chair that supports you. Now envision your project delivering results and creating impacts in the world. How does the product that you work on affect people's daily lives, zooming-in? How does it affect structures in society, zooming-out? Which virtues are relevant in this case? How does the product help or hinder people to cultivate these virtues? How might the product (better) support people to cultivate these virtues? You may also want to focus on the virtues that you need, working in this project. Do you need to cultivate courage, or self-control, or justice, or another virtue? Pick one or two. Now think of opportunities to cultivate these virtues. How can you try-out these virtues, taking small steps at first. Maybe there is somebody whom you admire, a moral exemplar, whom you can learn from?

Part III

Methods to integrate ethics

In this part (Chapters 16–21), we will first explore ways to combine the four ethical perspectives from Part II. Then, we will discuss different methods to integrate ethical reflection, inquiry, and deliberation in your projects. We will discuss Human-Centred Design, Value Sensitive Design, and Responsible Innovation. Please note that 'design' is meant here as a verb (*not* as a noun) and refers to a wide range of activities and processes: exploration, development, evaluation, and deployment. In these chapters, we will discuss, respectively, the organizing of iterative and collaborative processes; ways to take into account relevant stakeholders' values; and ways to promote anticipation, responsiveness, diversity, and reflexivity. There are practical recommendations and suggestions. However, because each project and each organization is unique, it will remain somewhat challenging to put these recommendations and suggestions into practice. The last chapter contains a series of short profiles of people whom I find admirable. I refer to them as *exemplars*. My suggestion is that you choose one or more of these people, watch videos of talks by them, listen to interviews with them, read their books, visit their websites, etc., to support you in cultivating virtues that are relevant for you and for your projects.

DOI: 10.1201/9781003088776-18

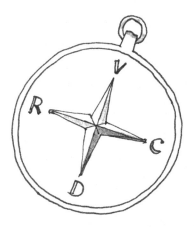

Chapter 16

Methods to 'do ethics' in your project

In Chapters 9, 11, 13, and 15, we discussed four traditions of moral philosophy. My proposal is to use these perspectives in parallel. Each one can supplement the others. I would guess that consequentialism is relatively accessible; people are used to assessing plusses and minuses. Deontology also comes relatively easily; people know about duties and codes of conduct and are familiar with the vocabulary of human rights. In contrast, relational ethics may be relatively new; it does, however, provide a valuable alternative to the other perspectives because it starts from different assumptions and asks different questions. Moreover, virtue ethics can play two key roles in your work: to understand how technologies can help or hinder people to

DOI: 10.1201/9781003088776-19

cultivate relevant virtues and thus promote or corrode people's flourishing; and to support you, as a professional, to cultivate virtues that you need in your projects.

Please beware, however, that these four perspectives *are* different. You do not want to confuse them. Like you would not want to confuse soccer, basketball, hockey, or water polo. These games are similar; they involve two opposing teams on either sides of a rectangle and the goal is to get a ball into a goal or hoop at the other end of the rectangle. But they are also different. They have different rules and you do not want to confuse them. You cannot use a hockey stick in basketball. And water polo happens in a pool and assumes that the players swim. Similarly, you do not want to confuse different ethical perspectives. Their goals may be similar, but their assumptions and questions are different. Maybe confusing ethical perspectives is like confusing billiards, baseball, golf, and cricket. They all involve balls. But they are very different. You do not want to throw a billiard ball in a baseball game. Or run laps between hole one and hole two on a golf course.

We have had good results with 'Rapid Ethical Deliberation workshops in which project team members were facilitated to look at their project and the innovation that they work on via these four ethical perspectives. Critically, they first describe the innovation in practical terms: which people will encounter it, how they will use it, which people are involved or implied, what it does for primary users and for others, on the short term and on the long term. The more detailed they make this description, the more detailed they can explore these four perspectives. They then can have a workshop, for example, of one or two hours, to go through the four perspectives.

Consequentialism helps to focus on human experiences, improving people's quality of life and minimizing people's suffering; to identify and evaluate positive and negative outcomes and impacts; to discuss and define system boundaries and 'externalities'; and to evaluate the distribution of plusses and minuses:

What are potential positive and negative outcomes or impacts of this project/innovation? You can think of impacts on ('internal') processes (zooming in) and on people's daily lives and society at large (zooming-out).

Where do you put your analysis' boundaries? What issues do you include or exclude?

How are positive and negative outcomes and impacts divided over different people or groups?

What could be unintended and undesirable outcomes or impacts ('side effects')?

Deontology helps to focus on human dignity and human autonomy, and on respecting, protecting, and helping to fulfil human rights; to identify duties and rights that are at stake; and to balance conflicting duties and rights, for example, by looking at different stakeholders' concerns regarding the project or innovation that you work on:

Does the organization that wants to implement this innovation have duties related to the innovation? If so, what are these duties? Maybe a *Code of Conduct*? Do *you* have any duties towards this organization? If so, what are these duties? Maybe a *Code of Ethics*?

Does this innovation impact on people's fundamental rights, for example, regarding human dignity, freedom, or equality? You can look at relevant (inter)national legislation.

Are there (informal) rules or legislations which the innovation needs to comply to?

Relational ethics helps to understand people as relational beings, as social and interdependent, and to understand how we are connected to nature, to focus on the ways in which technologies impinge on interactions between people, on distributions of power between people or groups, and on, for example, the effects on relationships and communication:

Which relationships, interactions, or collaborations would be affected by this innovation? You can think of interactions between people who use the innovation and of interactions between them and those who are affected by it. For example, between police officers who collaborate, and between a police officer and a citizen, respectively.

In what ways could the qualities of these interactions change? For better or for worse?

How could these changes affect power differences, communication, empathy, or care?

Virtue ethics helps to reflect on your project's outcomes' effects on society, on economic and social structures (zoom-out), and on people's abilities to cultivate specific virtues (zoom-in), to flourish and to live well together. In addition, virtue ethics can help *you* to cultivate relevant professional virtues, which you would need to cultivate and exercise in your projects:

Which virtues are at stake, or at risk, when people use this innovation? Think of both the people who use the innovation and the people affected by it. For example, self-control, empathy, or civility. How can the innovation help or hinder people to cultivate these virtues?

How might you modify the innovation so that it can (better) help people cultivate relevant virtues, to find appropriate means? On the level of society, how can the innovation help to promote justice, for example, in institutions, and promote people's flourishing?

Which virtues would *you* need to cultivate in this project? Maybe justice, courage, honesty, or humility? Maybe curiosity or creativity or collaboration?

Typically, I have found that one Rapid Ethical Deliberation workshop does not give all the answers you want. It needs to be part of an iterative and collaborative process; in which there is more than one workshop, and different stakeholders are involved. Even one workshop can, however, have immediate benefits. It can help the people involved in several ways: to envision the innovation they work on in practical terms; to look at it from diverse normative, ethical perspectives; to look at their project with fresh eyes, and to think about it creatively and strategically; and to articulate questions and actions to move their project forward.

For each project, there can be a different emphasis on one or more ethical perspectives. One perspective may be especially relevant for one project, whereas another perspective is more relevant for another. We can use some concepts from moral psychology to further explore this. In his book *Moral Psychology*, Mark Alfano discusses key concepts like patiency, agency, sociality, reflexivity, and temporality. Patiency refers to people's suffering or enjoyment of things that happen to them. Now, suppose that your project requires you to reflect on the ways in which the innovation you work on may impact on people (*patiency*), also over the course of time (*temporality*), then you may want to use a consequentialist perspective. Alternatively, if your project impinges on people's *agency*, their capacities to act in the world, or people's *reflexivity*, their capacities to understand themselves, you may need deontology. For projects that deal with technologies that bear on *sociality* and *temporality*, people's capacities to interact with each other, and how these may grow or languish over time, relational ethics can be useful. Finally, for projects that deal with *agency* and *sociality*, people's capacities to live together well, and with considering broader, societal implications, also over the course of time, you can turn to virtue ethics.

A moral compass

When talking about ethics, people sometimes use the term *moral compass*. That can be a helpful metaphor. It is, however, key to understand what we mean with that. Do we refer to something that tells us where to turn left or right? That sounds more like a satellite navigation system in a car. That would be handy. If something tells you exactly what to do at each junction. You would know the right thing to do in each situation. Or would it?

A compass, however, works differently. A compass only shows you where the magnetic North is. It does not tell you where to go left or right. A compass does not even tell you where you are. In order to navigate with a compass, you need three more things. First, you need a map. Typically you align the compass needle's North to the map's North. Second, you need to know where on the map you are. For that you will need to look at the terrain and use your perception and your memory of your journey so far. That mountain top you see over there must be this mountain top on the map. And the village you walked through an hour ago is here on the map. Now you can guess where you are on the map. Third, you need to have a goal to travel to, or at least a direction. We can translate the compass metaphor back to ethics. You need three things together with your moral compass. You need a moral map: a model of the situation that you are currently in. A moral map can show you swamps that you want to avoid, rivers that you would need to traverse, and sights that you do want to visit. You also need to know where you are on the map. You will need to look at the moral terrain. What is happening around you? You may want to use your perception to sense the situation you are in. Which people are involved? What are their values and concerns? Third, you need a goal or a direction. Maybe a specific goal. Maybe a general direction. What values do you want to protect and promote: justice, freedom fairness, conviviality, transparency? What kind of future do you want to help to create?

Curiosity and creativity

There is one more observation that I would like to share before we dive into methods to integrate ethics in your projects in the next chapters. Many people, myself included, have a tendency to think in terms of opposing values and interests. You are *for* something or you are *against* it.

During the Covid-19 pandemic, for example, concerns for public health and concerns for the economy were often thought of and presented as opposites. If you talk about promoting the economy, you are criticized for not caring about people's health. Or the other way around: if you talk about protecting people's health through lock-down measures, you are criticized for wanting to ruin people's businesses and livelihoods. As if protecting people's health cannot go together with protecting people's livelihoods. Similarly, people tend to think of security camera in public spaces as either good, because they help to prevent crime, or as evil, because they invade on people's privacy. As if we are unable to combine concerns for both safety and privacy, for example, by following a privacy-by-design process.

Curiosity and creativity are integral parts of design and innovation; they can help to combine different concerns and values. Take the example of recreational tents. In the 1960s, such tents were made of cotton, cut in rectangular or triangular pieces, supported by straight metal poles. These tents were rather heavy to carry around and rather small inside. If you wanted a larger volume, the tent became even heavier. And, conversely, a tent with less weight would become even smaller. Large volume and low weight were opposites. Enter curiosity and creativity. People explored new materials and new structures and came up with sphere-shaped lightweight tents, using synthetic, waterproof fabric, and flexible, fiberglass poles. Although there is tension between large volume and low weight, they *can* be combined. The key is to understand the requirements, depart from established notions, and explore new possibilities. (Surely, there are tensions that cannot be 'solved' by curiosity or creativity, but let us, if only temporarily, as an experiment, postpone judgement and promote curiosity and creativity.)

This is also why I discussed ethics in relation to other fields of knowledge (Chapter 2). Doing ethics, for me, is different from natural or social science, it is not about *what is*, but about what *could be* or what *ought to be*. It is more akin to design and engineering, which deal not with describing current states of affairs, but with envisioning and creating possible, desirable futures.

In the following three chapters, I will discuss three fields of practice: Human-Centred Design (HCD), Value Sensitive Design (VSD), and Responsible Innovation (RI). These fields share several key orientations and commitments. In order to avoid repetition, I will highlight specific elements in each chapter (while acknowledging that these are relevant also in other fields): ways to organize *iterative* and *collaborative processes* in the chapter on HCD; taking into account relevant stakeholders' *values* and *societal interests* in the chapter on VSD; and topics like *societal engagement, anticipation, responsiveness, diversity,* and *reflexivity* in the chapter on RI.

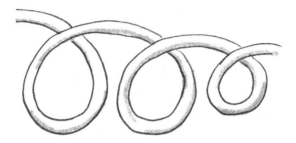

Chapter 17

Human-Centred Design

You have probably heard of 'design thinking'. Maybe an image springs to mind of a handful of people in a conference room, writing and drawing on pieces of coloured paper, sticking them on the walls. Maybe that gives you pleasant feelings. Maybe not. What I would like to do is talk about the *process* behind the surface of such an activity. I would like to discuss two ingredients that I believe are key to the process of design thinking: *iteration* and *collaboration*.

Iteration

In Chapter 2, I proposed that people in design and innovation can follow the logic of *abduction*; they start with envisioning a

DOI: 10.1201/9781003088776-20

desirable, future result, and then reason back to what is needed to realize that result. Design scholars like Kees Dorst, professor of Transdisciplinary Innovation at University of Technology Sidney, have suggested that we need to organize iterative processes that enable the people involved to move back and forth between problem-setting and solution-finding. Such an iterative process is especially needed when we want to address and cope with *wicked problems*: problems that cannot be clearly defined at the start of a project and that do not have one obvious best solution. Many of the large and complex challenges that we currently face are *wicked problems*. Polarization in society, for example, has many causes, and there is no single or simple solution. Diverse and partial solutions would be needed, and it is not immediately obvious where to start and how to proceed and develop potential solutions. There is no silver bullet.

Fortuitously, we can learn from pragmatist philosopher, John Dewey, whom we already met in Chapter 1. He proposed that the creation of knowledge needs to start with people's practices and experiences and needs to lead to knowledge that can be applied in practical situations and can help people ameliorate their circumstances. He advocated moving back and forth between engaging with practical situations and theorizing about these situations. He proposed an iterative and participatory process of inquiry, which consists of five phases. These phases do not need to be executed in any particular order, as they constitute an iterative process.

The first two phases consist of *exploring* and *defining* the problem. In a project, especially at the start, you can find yourself in an 'indeterminate situation', a specific and practical situation that you find problematic. But you cannot yet articulate the problem exactly. Dewey stressed that personal experiences and subjective questions can be key drivers for the inquiry process. *What bothers you in this situation? What do you feel needs to be solved?* As a team, you need to formulate a provisional problem definition, which you can later refine, in subsequent iterations. Wording and framing are important here. The way in which the problem is conceived of determines the type of solutions that you will entertain or dismiss. Very likely, the perspectives of various people, with diverse backgrounds, are needed to explore the problem from various angles and to define the problem in multi-facetted terms. Furthermore, there are moral and political dimensions to this: *Which perspectives are privileged or foregrounded, and Which perspectives are downplayed or ignored?*

The third phase is concerned with *perceiving* the problem and *exploring* possible solutions. This involves moving back and forth between problem-setting and solution-finding. Dewey proposed that, to better understand the problem, you can use your perception, your capacities to see, hear, touch, smell, and taste current situations (what is the case); and that, to explore and envision potential alternative situations and solutions, you can use your capacities for imagination (what could be). These processes feed off one another. Perceiving the problem more precisely helps you to envision more concrete solutions, just like the envisioning of concrete solutions helps you to perceive the problem more accurately. This process can involve using your imagination to envision and evaluate moral issues ('dramatic rehearsal', in Dewey's words). You can do exercises in which project team members, possibly together with putative users or other stakeholders, enact current, problematic situations, and envision alternative, desirable situations. *How does this situation feel? How is this solution better than the current situation?* Questions like this are also moral and political: *Which types of experiences or reflections are permitted ('in scope'), and Which are rejected ('out of scope')?*

Phases four and five involve the *creation* and *evaluation* of different possible solutions. Dewey would advise not to embrace one particular solution too quickly. The relationships between the tentatively defined problem and different possible solutions will need to be evaluated carefully. These relationships are like working hypotheses. Ideally, project team members can explore and define the scope and boundaries of a project and critically discuss means and ends: *What exactly is the problem that we are trying to solve? And how would this particular solution help to do that? Has the problem definition changed over time? Do we need to modify it, based on new insights?* Such questions are moral and political in that they involve privileging or disregarding certain values or concerns. Additionally, thinking about means and ends requires systems thinking and dialogues about the scope and boundaries of the project at hand.

Moreover, these phases involve trying out solutions in practice, for example, in experiments with potential users or customers. In the spirit of pragmatism, such experiments are not about finding a positive answer ('yes, it works'), but about testing the relationship between means and ends: *How does this product or system bring us closer to the goal we want to achieve?* In this phase, the project becomes more real and stakes get higher. People become interested. It is important to stay critical of the project in these phases. In the *theory* of innovation management, it is possible at various 'stage gates' to abort a project, for example, because the results do not meet the expectations. In *practice*, however, people find it hard to stop. One of the first projects I worked on was an exception. The commissioner had asked me to create several paper prototypes for some telecom service and to organize a series of focus groups to evaluate people's reactions. I found that none of the participants was able to mention any benefits of this new service. After my presentation, the commissioner made a phone call to abort the project, there and then. I see this as an example of a *successful* project; cancelling a project for good reasons (and avoiding further costs).

Typically, people in project management roles prefer a linear process. Unless they have adopted an iterative approach, which is increasingly the case, for example, with agile approaches. Still, many projects are managed in a linear fashion: like an arrow that goes from left to right, or like a waterfall that flows from top left to bottom right. Asking questions about the problem statement may then feel like going back. Or like going in a circle. *'We did discuss that at the project's start'*. *'We decided this six months ago'*. This can be especially pertinent in large, multi-year projects. These often use a *Description of Work* to steer the project. This document can, however, be five years old; when there was one year between writing the proposal and starting the project and when we are already four years into the project. In such a case, we need to look up on, say, page 46, what we are expected to deliver in month 48.

Alternatively, an iterative approach allows the people involved to combine both moving forward and gaining new insights and integrating these in the project, for example, by revisiting issues from the past and shedding new light on them. For several projects, I made drawings of the project's structure and planning. I refrained, obviously, from drawing an arrow or waterfall. But how could I draw an iterative process? With a circle? I've found that drawing a circle gives the impression of 'the project getting nowhere'. Instead, I have drawn something like a spiral that becomes more narrow and focused from left to right. This suggests a forward motion and progress over time: scoping the problem-space and clarifying the solution-space. In addition, the circle shapes in the spiral suggest moving back and forth between creating and evaluating different solutions.

Coming back to the image of people writing and drawing on coloured pieces of paper. What I found in many creative sessions or workshops is the need for clarity about two processes that are both needed in creativity: *divergence* and *convergence*. In a creative process, people need moments to explore and diverge. This is what many people associate with creativity. But on other moments, they need to make decisions and converge. This was a key message of Jan Buijs (1948–2015), who was professor in Management of Innovation at the school for Industrial Design Engineering at Delft University of Technology. (Jan supervised my PhD, of which I have fond memories.)

It is, therefore, critical to manage these different moments and impulses: for divergence and for convergence. When do we need to explore and diverge? When do we need to make decisions and converge? Likewise, in project meetings, it can be good to clarify and make conscious decisions either to discuss the scope of the problem or to discuss directions for possible solutions. It can be confusing, and indeed unproductive, when these processes intermingle, or when one type of activity is privileged, possibly unintentionally, over another type of activity. Notably, these processes can be combined productively. It can be productive, for example, to make (clear, well-reasoned) decisions about directions we want to explore; or to explore (creative, surprising) criteria that can help us make decisions.

Over the years, I have observed a tendency, in projects, to focus on means, on getting things working, rather than to discuss goals that we find worth pursuing. Especially when the latter involves uneasy questions, like: *With this new information, how can we modify the original problem statement?* Or questions related to problem-setting, like: *For which party or people is this a problem? In what sense?* Or related to solution-finding, like: *In which directions are we (not) allowed to look for solutions? Which party's interests are served with this type of solution?* We will visit similar questions in the next chapter, on Value Sensitive Design.

Let me illustrate the organizing of an iterative process with the TA2 project (already mentioned). In this project, we developed five telecom and multimedia applications that were meant to connect family and friends, so that they can experience togetherness. For each application, we moved through the following steps: first scenarios in the form of short written descriptions; then scenarios in the form of storyboards; then demonstrators in the form of mock-ups (partly functional), which were used in laboratory tests; and then demonstrators in the form of working prototypes, which were brought to people's homes for trials of several weeks. For each step, we invited people from different potential user groups: for focus groups (with storyboards), for user tests (with mock-ups), and for trials in their homes (with prototypes). This iterative process, in collaboration with potential users, helped the project team members, both on the content level and on the process level. On the content level, it helped them to express, discuss, and prioritize ideas; and to move from storyboards, to mock-ups (user tests), and to prototypes. On the process level, it helped them to empathize with potential users; to integrate different tasks within the project, collaborate, and deliver tangible results; and to bring focus and realism in the project. Communicating with people who interact with the application you work on, can have rather humbling effects.

Collaboration

Interestingly, there is a standard for *Human-Centred Design for Interactive Systems*, which describes the following key principles: to start with an explicit understanding of prospective users, and their contexts and tasks, and to view their experiences holistically, not only in terms of usability, but also in terms of their emotions and aspirations; to involve prospective users throughout the process of design and development and in timely and iterative evaluations; to organize an iterative process; and to organize multidisciplinary teamwork.

We already discussed the need to focus on the experiences of prospective users (in Chapters 9, 11, 13, and 15) and the need to organize an iterative process. We can, therefore, now turn to the need to organize collaboration. We will look at collaboration on two levels: between people who work in a project team; and between them and prospective users of the system that they work on, and other people who will be affected by this system.

Let us first look at collaboration in a project team. In order to understand the problem that any project will focus on, we will need people with different fields of expertise. Which fields spring to mind? Computer science and data science, software development and software engineering, to develop systems that actually work. Cognitive psychology, social psychology, and user experience design, to understand prospective users, and their contexts and experiences (zooming-in). Sociology, ethnography, economics, and political theory, to understand wider societal contexts (zooming-out). And business administration and finance, to develop sustainable business models. And, people who can manage the project and foster collaboration between the people involved, both within the project team and between the team and various stakeholders.

Maybe you have experiences with collaboration between people from different disciplines. I have found that this can be challenging. Putting them in a room, around a table, does not immediately make them collaborate effectively and efficiently. Very likely, collaboration will need to be facilitated, with, for example, methods for *transdisciplinary innovation*, a topic we will discuss in Chapter 19 on Responsible Innovation.

The other type of collaboration we need to discuss is between project team members and prospective users of the system they work on and others who will be affected by it. There is a range of methods available to facilitate this type of collaboration. Participatory design is a classic method, with roots in Scandinavia in the 1970s. Its aim has been to empower prospective users of computer systems to have critical roles in designing these systems. It was closely related to the workers' unions and political action. This aim resonates in *Co-design* or *Co-creation*, but typically with less explicit political goals. Participatory design is concerned with enabling prospective users to participate actively in research, design, development, evaluation, and deployment. Collaboration is also possible in the other direction; project team members can engage in *Empathic Design* or *Contextual Design* in order to move towards prospective users and, to some extent, participate in their contexts. Such methods draw inspiration from social science, notably from ethnography and other types of fieldwork and facilitate project team members to better understand and empathize with prospective users.

What is critical, in these methods, is to understand that users are *experts*. Probably not experts in science or engineering. But experts of their own contexts, tasks, and experiences.

In abstract terms, these methods aim to bridge two gaps: the gap between project team members and prospective users, either by enabling users to move towards the project and participate in it or by enabling team members to move towards users and empathize with them; and the gap between concerns for studying and understanding current (problematic) situations (a research orientation; 'what is'); and concerns for envisioning alternative (desirable) situations (a design orientation; 'would could be').

Furthermore, these methods require commitments to both curiosity, research, and learning and to creativity, design, and productivity. Team members and prospective users need to find ways to engage in collaborative curiosity and collaborative creativity. I have found that this can require effort and patience. Otherwise, participation stays at the surface level.

I have participated in numerous workshops and meetings that were less about curiosity and creativity and more like ticking boxes. We did a focus group with customers. We evaluated the prototype with users. But we, the project team members, stayed within our own frames of reference. Clung to our ideas. And learned very little. The 'users' did not participate a lot. And we did not empathize a lot with those 'users'.

If you stick too much to your interview format, you dampen curiosity. If you stick too much to your workshop format, you hamper creativity.

Let me give an example of collaboration. In a project for the new business development department of a telecom operator, we were asked to explore future needs and related mobile services. This was in 2007, when the first iPhone was introduced. We organized two tracks in this project: in one track, we did Delphi-style workshops with various experts, to hear their views and arguments; in another track, we organized three co-creation workshops with a total of 50 children of 7–10 years old. Children of that age are relatively open to curiosity and creativity. Notably, we invited the company's employees' children; not only for practical reasons, but also for strategic reasons. We wanted our project to be not only about content, about mobile services, but also about organizational change. Our intention was not only to facilitate these children's curiosity and creativity, but also to spark curiosity and creativity in their parents. We took care to make the children feel safe; their parents were nearby, in the company restaurant, and people with experience with working with children were facilitating groups of 4–5 children. In the workshops, we started with storytelling, involving wizards, time-travel, or stranding on a desert island, to explore and articulate (future) needs, and then children used all sorts of scrap materials, like carton, plastic, and metal, to create mock-ups of future mobile services. At the end, the children presented their ideas. The project was successful: the client was happy with the surprising and innovative ideas; and they also noticed more curiosity and creativity in their employees, the children's parents.

Being aware of the challenges and limitations of the methods mentioned above will help you to apply them appropriately. You can cultivate reflexivity, to become more aware of your own role in these various interactions, which will enable you to cultivate curiosity and creativity. You will need to balance your own ideas with other people's ideas. You can consciously decide to sometimes stick to your own ideas and at other times to privilege other people's ideas.

In closing, I would like to remark that I find the term Human-Centred Design slightly problematic. First, we need to ask: Which people will we focus on? And whose concerns will we prioritize? Those of customers or users? Or of other people, who are affected by the innovation? We need to make explicit our moral and political commitments. Moreover, I would like to remark that I use the term *not* to refer to an exclusive focus on people at the expense of other animals, plants, our natural environment or our planet. Rather, I understand it as an effort to promote long-term wellbeing of people as part of nature: as taking care for people, for non-human animals, for life on earth, for the earth as an eco-system. Terms like *social justice oriented design* or *sustainable development* could be more appropriate.

How do you see iteration and participation in the projects that you work on? Do you and the people involved make room to question the problem statement before you dive into creating solutions? Do you organize iterations, deviate from linear project management, and revisit decisions? Do you and the others involved, collaborate with potential users or other stakeholders? Do you make room at the table for a variety of people with diverse backgrounds and disciplines? And, if not, what is holding you back? What can you do, to promote iteration and collaboration?

Chapter 18

Value Sensitive Design

Many projects are driven by people's values. The people working on a project look at the world, believe that something is lacking or needs to be improved and orient their project towards goals that they believe to be worthwhile. Projects may be motivated by all sorts of values: moral or political values, like justice, fairness or democracy; or economic or business values, like efficiency, usability, or profitability. We can understand design and innovation projects as attempts to create value, in terms of wellbeing or economics (see Chapter 4). In such projects, people attempt to put specific values into systems, in the hope that, when these systems are deployed and used in practice, these values are realized and become material. These systems offer

DOI: 10.1201/9781003088776-21

specific opportunities and specific constraints to the people who interact with them. The values become, as it were, embedded in these systems and the processes associated with them, and these systems shape people's behaviours (see Chapter 3). Let us further explore what we mean by values.

Values

Value Sensitive Design (VSD) puts values centre stage. It is a method that enables people to explicitly discuss, explore, and negotiate values, carefully and systematically. VSD aims to enable diverse stakeholders to express their values and to combine these productively during the innovation process. VSD has much in common with HCD and has, for example, borrowed from its methods.

Batya Friedman, professor at the University of Washington, and others pioneered VSD in the context of computer science and interaction design. They proposed to focus on values like: human welfare; ownership and property; privacy; freedom from bias; universal usability; trust; autonomy; informed consent; accountability; courtesy; identity; calmness; and environmental sustainability. This is, of course, not the only list of values. For the design and deployment of algorithms, for example, relevant values would be: accuracy, autonomy of both human decision makers and data subjects (citizens), privacy, fairness, property, accountability, and transparency.

It can be worthwhile to distinguish between instrumental and intrinsic values. Instrumental values help to realize other values; their value is in their contribution to these other values. In contrast, intrinsic values have value in and of themselves. Accuracy, for example, can contribute to transparency and together they can contribute to fairness. In this example, accuracy and transparency are instrumental values, whereas fairness is thought of as having intrinsic value.

For a more wide-ranging list of values, developed independently of the context of technology or innovation, we can turn to moral philosopher William Frankena (1908–1994). He suggested the following values, grouped into six broad categories: beauty, harmony, proportion, aesthetic experience; moral disposition, virtue, pleasure, happiness, contentment; truth, knowledge, understanding, wisdom, honour, esteem; life, health, strength, peace, security; love, affection, friendship, cooperation; and power, achievement, freedom, adventure, novelty.

For your specific project, you can look at a list like this and select those values that are most relevant for your project and create a more manageable and focused shortlist of values.

I have organized workshops that can be understood as a scaled-down version of VSD. In these workshops, we helped clients to integrate ethics in their projects, which involved the collection of data and the creation and deployment of algorithms. For example, to monitor the effects of social welfare services in a large city and to create suggestions for follow-up actions to promote these services' effectiveness. Or to monitor people's viewing habits in a public broadcaster's mobile app and to create suggestions that promote more diversity in their viewing habits. In each workshop, we worked with a 3×3 matrix. The matrix' three rows represent the process from data to application: data selection and collection; data processing and modelling; and the algorithm's output and using it. The matrix has three (or more) columns, to represent values like: human autonomy, fairness, transparency, or privacy. First, we clarified how the application works and discussed which values are relevant. Then we selected three (or more) cells in this matrix to focus our discussion on. For example, fairness in selecting data sources and collecting data, transparency in modelling and creating the algorithm, and users' autonomy in using the algorithm's output. For each cell, we discussed issues and potential solutions for 20–30 minutes.

A complete version of VSD consists of three types of investigations, which can be combined in an iterative process: empirical, conceptual, and technological investigations.

Empirical investigations involve studying the values that are at play in a particular project, for example, by conducting interviews or workshops with relevant stakeholders. Proponents of VSD have borrowed from HCD and created methods like *Value Scenarios*, which are short narratives that project team members create in order to express how the system they work on may impinge on key values. These investigations result in (tentative) requirements for the system that is being developed. They also result in a better understanding of how different designs may affect different values. Design A may be good for value X, but not so good for value Y; whereas Design B is good for both X and Y, but not so good for value Z.

Conceptual investigations involve further studies of values that are at play; for example, to better understand relationships or hierarchies between different values. Which values are instrumental to realize other values? Which values are, in this case, more important than others? One can draw diagrams, with 'ladders', to explore how different technical requirements and solutions enable (or restrict) different behaviours, and how these behaviours, in turn, enable (or restrict) the realization of different values. The 'ladders' connect technology, via behaviours, to values. Typically, conceptual investigations also involve the exploration of tensions or trade-offs between values. We did already meet value tensions in the trade-offs between volume and weight of tents and between security and privacy of surveillance cameras. It is worthwhile to also look at legislation, organizational practices, or social norms that pertain to the values at stake.

Technical investigations involve the actual design of a system. Based on findings from the empirical and conceptual investigations, we can generate functional requirements and design options that aim to meet the most relevant values of the most relevant stakeholders. Here, it is important to keep in mind that VSD is not a hard science. There are no formulas to create brilliant design solutions, nor are there formulas to assess which option is the absolute best.

Now, there are several challenges with VSD. First, it is critical to distinguish *values* from, for example, *interests* or *preferences*. I understand *values* as referring to concepts that we have reason to value, for example, access to education or to health care, or capabilities to engage in friendships. They are often common goods, which are recognized broadly, although, of course, people can differ in the relative weights they give to different values. Conversely, I understand *interests* as people's stakes in a particular matter. One person's interests can conflict with other persons' interests. Or collectively, one group's interests can conflict with another group's interests. One group has an interest in paying less tax, whereas another group has an interest in public spending on education or health care, which will depend upon collecting taxes. And there are *preferences*, which I view as similar to interests, but with less moral or political significance, for example, how one prefers to spend one's leisure time. Some may want to spend time with friends in a park, have a picnic and play music, whereas others may want to be in the park in order to enjoy its beauty in quietness. In this example, people's preferences may coexist without problems, or a conflict may arise, depending on the size of the park and the willingness and abilities of the people involved to modify their behaviours. I would propose that dealing with different organizations' or people's interests and preferences would fall under the header of *stakeholder management* or *project management*, not under the header of VSD.

Another challenge relates to *value tensions*. How can we deal with different views or weights that different organizations or people give to different values? One way to deal with this is to explore ways to combine seemingly conflicting values. One can apply methods, for example, from HCD, to spark curiosity and creativity, and thereby enable participants to move beyond dichotomies and to create solutions that satisfy conflicting values, like the example of the large and lightweight tent.

Furthermore, it is critical to appreciate that VSD, like HCD and Responsible Innovation (see next chapter) are *social* processes, in which a range of moral and political issues are always at play, often implicitly. Look at statements like 'the most relevant stakeholders' and 'the most relevant values'. Who gets to decide which stakeholders are 'most relevant', and how do we determine which values are 'most relevant'? It can be challenging, practically and logistically, but also morally and politically, to select and invite stakeholders. *Who will you invite and include? And who to will you not invite and exclude? Do you invite A because you have already collaborated agreeably with them? Do you not invite B because one of their people said something unpleasant about one of your people?* Likewise, it can be challenging to determine which values to include in your analysis. And which to exclude, more or less consciously. Values like *usability* and *privacy* are often perceived as relevant in technology and innovation, whereas values like *wellbeing* or *fairness* can be perceived as difficult or contested and may, therefore, be less welcome.

Moreover, there are issues of translation. There can be differences between *intended* values, which, during design, become *embedded* in a technological system (or not) and which, during usage, are *realized* (or not). Intentions can get lost during design and usage. The translations involved in design to usage are messy and complex, not simple, instrumental, or deterministic. We saw examples of this in Chapter 3, where we discussed how people shape technologies and how these technologies in turn shape people's behaviours.

Another challenge is that discussion about values can remain relatively abstract. The people involved can have different ideas in their heads about the innovation that they are working on. Hence, their ideas of values at play can remain relatively abstract. This can be remedied by clarifying, early on, what the innovation will look like practically, and what it will do for people. Here is another challenge, which often manifests as a pitfall. The pitfall to talk *about* other people, make assumptions and relay hearsay, rather than talk *with* these people: listen to their experiences and learn from them. Here, VSD can borrow from methods for HCD, to envision the innovation in practical terms, and to invite potential users to talk about their experiences.

Capability-Oriented Innovation

In my experience, I have often found that people find values rather abstract. Alternatively, we can work with human capabilities. We can orient our innovation projects towards enabling people to develop relevant human capabilities, to contribute to creating a just society and promote human flourishing. This approach borrows from the capability approach (Chapter 4) and from virtue ethics (Chapter 15). Examples of human capabilities are: to live with good health, to live in security, to move around freely, to express and share ideas, and to participate in societal and political processes. Several people have similarly proposed *Capability Driven Design* or *Capability Sensitive Design.*

Suppose that you work in a project that aims to promote people's health, by creating an app that enables people to cultivate healthy habits regarding diet and exercise. You can articulate that goal as: enabling people to develop their capabilities to live healthily. Preferably, you do this at the project's start. This helps you and the others involved to orient the project towards practical goals and to avoid focusing too much on the app that you are working on ('only a means').

Imagine that this project involves collaboration between a university spin-off start-up company, with knowledge on behaviour change in the health domain; an insurance company, that has access to a large group of potential users; and a software company, which builds the app, where you work.

Each partner may have a (slightly) different view on the project's goal and on the means to achieve that goal: the app. The university spin-off people focus on their models for behaviour change and want to collect data for their research. The insurance company people want to offer the app as a premium service, to help their customers to stay healthy and cut health care expenditures.

Their different views may lead to different functional requirements for the app. In such a case, it may be worthwhile to promote a shared understanding of the project's goals. This can be done, for example, through conversations about the project's outcomes, in terms of how the app can support people to develop relevant capabilities. It is worthwhile to spend some time on this early on in the project, to prevent misunderstandings and conflicts later. Different people in the project consortium can express their ideas on how they want to enable people to extend their capabilities to live healthily with regards to diet and exercise. They can also refer to other capabilities, like autonomy or self-determination. Such conversations can enable them to make choices about the ways in which data are (not) collected and (not) shared with third parties, for example.

Moreover, conversations like this can help to solve a key problem in innovation: the adoption by various relevant actors of a project's outcomes. Without adoption, the innovation ends up in what some refer to as the *valley of death*. In this regard, we need to distinguish between: a project's *outputs* (or deliverables); the *outcomes* it aims to realize, during and after the project; and the *impacts* that it aims to contribute to, eventually.

First, we need to turn a project's *outputs*, say, a prototype of a health app, into *outcomes*, say, the successful adoption and deployment of this prototype by relevant actors. Then, further efforts are needed to create larger *impacts*. This would typically involve the creation of an innovation ecosystem of various actors and stakeholders, which can turn outcomes into desired impacts. In this case: improving people's health and reducing health care expenditures.

We can apply this *Capability-Oriented Innovation* approach in the following basic steps:

Discuss and clarify the project's goal in terms of enabling people to extend one or more *human capabilities* and, thus, to promote human flourishing. The system, product, and service that the project aims to deliver functions then as *means* towards that goal.

Invite partners and relevant actors to express their views on these human capabilities. The capabilities that they aim to support through this project can then be translated into requirements for the system, product, or service that the project aims to deliver.

Create and maintain an innovation ecosystem that enables not only the design phases, but also implementation and deployment. New partners may be needed later on. Initial partners may no longer be needed. This requires sustainable business models for collaboration.

From the chapter on HCD (Chapter 17), you will recall the need to involve partners, actors, or stakeholders as timely as possible. Otherwise, the project may suffer from 'not invented here'; they will find it difficult to accept and embrace other people's ideas and findings. Also from HCD, we know that the involvement of prospective users and other stakeholders is critical. In the health app example, this could be people who want to use this app, or relevant professionals, for example, medical doctors, general practitioners, or dieticians.

In closing, let me focus on one particular challenge of VSD. Values can *change* over time, due to changes both in technology and in society. Ibo van de Poel, Antoni van Leeuwenhoek professor in Ethics and Technology at Delft University of Technology, has studied this challenge. He proposed that since new technologies offer new opportunities and enable new practices, they can very well lead to new practices and new experiences, which pose new obligations and dilemmas. He discussed the emergence of new values, changes in different values' relevance for different applications, changes in how various values are understood, their relative importance, and their translation into design requirements. We will come back to this in the next chapter when we discuss anticipation and responsiveness.

Which values are at stake in the project that you work on? Which stakeholders do you make room for in your project? Which values are important for them? What can you do to learn more about less obvious stakeholders and their values? You may want to explore ways to translate the project's outputs into outcomes, and into impacts in society. This may involve creating an innovation ecosystem. How does the project that you work on contribute to creating a just society, in which people can extend their capabilities and flourish? How can you steer the project so that its outcomes can enable people to develop relevant human capabilities?

Chapter 19

Responsible Innovation

The term *Responsible Innovation* (RI) refers to the ambition to (better) align innovation projects to needs and concerns in society; notably with, for example, requirements for sustainability and societal goods, typically, in close collaboration with, for example, Civil Society Organizations. The term was pioneered by the European Commission but is currently used more broadly. I would like to supplement the meaning of RI and use it to refer to the design and deployment of social and technological innovations that help to create conditions in which people can extend relevant capabilities and cultivate relevant virtues, so they can live well together and flourish.

DOI: 10.1201/9781003088776-22

Two key ingredients of RI are: *societal engagement*, that is the involvement of societal actors, stakeholders, and citizens in the innovation process, to learn about their needs and concerns, and to take these into account during design and deployment; and *ethical deliberation*, that is the organizing of processes in which the people involved can identify ethical issues, invite diverse views, investigate issues, and make choices responsibly. We already discussed ways to integrate ethical reflection, inquiry, and deliberation in projects, so we can currently focus on societal engagement.

Societal engagement

In the field of innovation management, societal engagement can be understood in terms of a *quadruple helix*. This image refers to collaboration between industry, academia, government, and societal actors, like a Non-Governmental Organization that represents some societal concern. That can indeed be very worthwhile. Societal engagement, however, can also be a hassle. You will need to select and invite organizations to collaborate with. This involves moral and political considerations. Which organizations do you include or exclude? And once the people are in a room, around a table, you will need to facilitate communication and collaboration. There will be tensions between participants. You will need to take into account their different concerns and integrate these in meaningful ways.

It can be hard to convince yourself or others that this hassle is worthwhile. Based on a literature survey, I created an overview of benefits and challenges of societal engagement. You can use elements from this overview to discuss the pros and cons of societal engagement.

The involvement of societal stakeholders can bring benefits like the following: *Outside-in orientation*, a better understanding of needs and concerns in society, for example, the problems that people encounter, which can help to design better solutions; *Legitimacy*, for the project, for example, to enable citizens to voice their concerns, so that these can be taken into account; *Alignment*, to better align the organization's strategy to concerns in society, for example, the needs of specific citizens; *Collaboration,* to build relationships, for example, to co-create agendas for future actions and to collaborate with lead users; and *Clarity*, within the project, for example, about the scoping of the problem that the project will work on, and hence, promote better making of decisions and finding of solutions.

Conversely, societal engagement can bring the following (perceived) challenges: *Effort,* it can require time and resources and often also expertise and commitment to do this well; *Complexity*, as in 'too many different viewpoints', and with it the perception of the risk of having less control (but it can be beneficial to know these viewpoints, rather than act as if they do not exist); *Slowdown,* the perception of the risk of slowing down the project (but this does not have to happen; societal engagement can actually help to speed up the project); *Expectations,* to manage the expectations of the organizations and people involved; and *Reputation,* for example, the risk that an interlocutor talks about sensitive issues to a journalist who is looking for a scoop. Please note that I put 'perceived' in front of these challenges. Some challenges are (partly) perceived (as I suggested with remarks in brackets) and can be dealt with relatively easily.

There are all sorts of ways to collaborate with societal actors, for example, with citizens or citizens' representatives, ranging from more creative methods, like a *world café* (where a relatively large group of people is divided into smaller groups, typically around tables, to explore topics and discuss issues), to more political methods, like *citizens' assemblies* (where citizens, typically randomly selected, are facilitated to collectively deliberate on important issues, with a specific political mandate). With such methods, it is critical to provide citizens with knowledge and to facilitate communication. When executed well, a citizens' assembly can lead to successful policies. In several countries in the EU, for example, citizens' assemblies have been organized to help create policies to reduce carbon emissions and combat the climate crisis.

Needless to say that the project's scope and budget will determine which method you choose and the scale at which you can execute it. Maybe two round table discussions with three relevant Civil Society Organizations are enough. Maybe you need a series of meetings with larger groups, maybe also with citizens, and maybe you need to iterate these over the course of the project.

An example of implementing societal engagement, from the organization in which I work, TNO. TNO was founded by national law in 1932 with the goal to support government and industry with applied research and innovation. It is organized in nine Units, each dedicated to a particular aspect of society (Industry; Healthy Living; Defence, Safety & Security; Buildings, Infrastructure & Maritime; Traffic & Transport; Circular Economy & Environment; Energy; ICT; and Strategic Analysis & Policy). Each Unit has a Strategy Advisory Council (SAC), with representatives from government, academia, industry, and society, as a way to better align the work of TNO with needs in society. I was part of a team that was invited to improve this alignment. We noticed that these SACs had relatively few people from Civic Society Organizations (and many from industry) and relatively few women (in 2017). We spoke to people in our organization about ways to improve this and supported them. This resulted in more inclusive and diverse SACs; the percentages both of Civic Society Organizations and of women doubled (in 2018).

Another way to understand RI is to discuss its key dimensions: anticipation, responsiveness, diversity, and reflexivity. I will discuss these in the next three sections; I will take anticipation and responsiveness together.

Anticipation and responsiveness

Anticipation refers to the exploration and evaluation of potential outcomes and wider impacts of the project that you work on: both desirable and undesirable impacts, and both direct (first-order) and indirect (higher-order) impacts. Anticipation will need to go hand in hand with responsiveness. After having explored potential outcomes, you will want to respond appropriately. You may want to steer your project towards more desirable outcomes. Or away from undesirable outcomes. I would like to propose that anticipation and responsiveness need to go hand in hand. We need to anticipate a projects' potential impacts and take appropriate actions based on these findings. Together, they enable effective learning. This can be done, for example, by organizing an iterative process (see Chapter 17).

When we talk about anticipating the undesirable impacts of innovations, we can think of multiple examples. DDT. Asbestos. Their devastating effects are well known (accumulation of toxic DDT in food chains; lung cancer in people who came in contact with asbestos). We can also think of innovations with mixed effects; some desirable, some undesirable. Combustion engines. Plastics. Social media. In and of themselves, combustion engines, plastics, and social media are not evil. But in practice, they have helped to shape larger systems that have effects that are more undesirable than desirable. The inventors of the combustion engine did not intend to cause a climate crisis by burning fossil fuel, nor did they intend to cause suburban sprawl and its social problems. The inventors of plastics did not intend to pollute the rivers and oceans with microparticles. Notably, Swedish engineer Sten Gustaf Thulin, who invented the plastic shopping bag in 1959, intended to do good for the environment. Plastic was better than paper bags, for which trees need to be chopped. He assumed that people would treat plastic shopping bags as durable and re-usable items.

There is an (apocryphal) story that can illustrate the mechanism of good intentions and unintended, undesirable effects. In the time of the British rule of colonial India, the British wanted to get rid of venomous cobras in Delhi and offered a bounty for every dead cobra. The intended effect was that people killed snakes for this reward. People, however, also began to breed cobras in order to claim their rewards. The government found out and stopped the reward program. The breeders then let their cobras go free, which were now worthless, which worsened the cobra plague. Officials could maybe have anticipated this effect if their system analysis had included variables and feedback loops for supply and demand and for people's motivations and behaviours.

Unintended, undesirable effects are especially and notoriously difficult to anticipate and mitigate when they involve not only direct or first-order effects, but also indirect or higher-order effects.

The time dimension is critical here, since higher-order effects typically happen over the course of time. One way to promote anticipation and responsiveness is to envision future scenarios or use cases, build demonstrators or prototypes, and organize experiments with them. These can be relatively small-scale, cheap, and quick; they are meant to experiment and learn. More elaborate studies can borrow from the tradition of (Constructive or Participatory) Technology Assessment.

Importantly, some higher-order effects will only materialize once systems are actually being deployed and used in people's daily lives. This means that experimenting and learning are not finished after the system is implemented. Ideally, experimenting and learning can be continued in some form during a system's deployment and usage.

In a recent article, we explored potential, unintended, undesirable effects of decision support systems, especially the effects that may happen over time. We gave the example of a system that a government agency uses to assess ('predict') the risks of fraudulent behaviour of citizens. Typically, such systems draw from multiple data sources and put 'orange flags' for specific people's names. We suggested that, despite good and legitimate intentions and despite careful and professional work processes, such systems can have unintended, undesirable effects. Let us look at one of the principles that the European Commission's High-Level Expert Group put forward: human autonomy. *It is critical that agents who use the system are able to combine the algorithm's output with their professional discretion. The extremes of slavishly following its output or of consistently neglecting its output will typically need to be avoided. This requires effective Human-Machine Teaming and Meaningful Human Control. Over time, as, for example, new features are added to the system, or as the system is applied to new domains or types of cases, however, the ways in which people use the system may drift. It may drift towards excessive automation, where too many tasks are delegated to the system, so that people no longer have meaningful control. The system then performs tasks that require human perception, discretion, and judgement. Or it may drift towards excessive human control, where people need to micro-manage the system, which can be very inefficient or which can lead to overstressed operators who make mistakes. The system may also drift away towards* too little computer automation, *where people perform too many routine tasks and start to make mistakes, because of reduced concentration. Or towards* too little human control, *where people learn to follow the algorithm's output unthinkingly and routinely click the OK button.*

Diversity, inclusion, and participation

I will discuss diversity in combination with inclusion and participation because I believe they are related. One way to promote diversity, inclusion, and participation is to collaborate with putative users and with relevant stakeholders. We already discussed this in Chapters 17 and 18.

Let me give two examples of enabling putative users to participate in the project, and to learn from them. One example is from a project back in 2004 (Freeband FRUX). We were a project team of ten researchers and developers, working on mobile telecom services, one for police officers, and one for informal carers. I proposed that each of us spend a day with a police officer, on the street, to better understand their contexts and tasks, and to visit informal carers, to better understand their contexts and tasks. People in management roles questioned what we would learn from these people. But we proceeded. And our encounters were very meaningful; in the course of the project team members kept referring to these encounters on many occasions. Another example is from John Thackara, from 1999, when he was leading a project that aimed to create

Internet-based services for the elderly. Thackara wrote the following about their first encounter with 'their' putative users:

'Someone said, "There are a lot of older people out there; let's see if we can find some and help them by giving them this Internet stuff in an easy-to-use format". So we went and found some older people and told them how we had come to help them with the Internet, and they said, "Piss off!" which is apparently how they say, in some long-lost dialect, "We don't need your patronising help, you designers. If you've come here to help us, you're wasting your time; we don't want to be helped, thanks just the same. Yet we do have some interesting observations to make about our daily lives, about our lifestyles, about our communication, and about all of their attendant dysfunctions. If you could kindly change your attitude and help us explore how we will live, then perhaps we can do something together". Or words to that effect'. (This is also an example of reflexivity; see next section.)

Another way to promote diversity, inclusion, and participation refers to the composition of the project team. Now, why would a project team's diverse composition be relevant?

In Chapter 17, we discussed the need to involve multiple disciplines in order to address and cope with *wicked problems*, for example, the need to transition from fossil fuels to sustainable energy, and the need to combat social and economic inequalities and injustices. We need people from diverse organizations and from different disciplines to study and understand these problems and to explore and develop possible solutions. Here, transdisciplinary innovation can help.

Transdisciplinary innovation goes beyond *multidisciplinary* and *interdisciplinary* approaches. In a multidisciplinary approach, people use their different disciplines to jointly look at a problem; a social psychologist and a computer scientist collaborate to study the role of attention in how people use a specific mobile app, for example. In an interdisciplinary approach, people combine their disciplines at the interface between these disciplines; a social psychologist, a computer scientist, and an economist study how companies use 'sticky' mobile apps to capture people's attention and how this aligns with a specific business model. In a *transdisciplinary* approach, people integrate their disciplines at the intersection of research, societal issues, and innovation; a social psychologist, a computer scientist, a policy maker, and an expert in public administration collaborate to study the above questions. Moreover, they can explore and propose solutions, for example, ways in which people can use 'personal data vaults', to selectively give specific organizations access to their data, or to form 'data commons', union-like organizations in which they can aggregate and share their data, while remaining ownership and upholding civic values.

The people involved in transdisciplinary innovation need to move outside and beyond their own disciplinary boundaries and let their fields of expertise interact with other people's fields of expertise. Innovation can then happen in the space between their disciplines.

While transdisciplinary innovation is direly needed, in coping with wicked problems, I have noticed that people can tend to underestimate the difficulties of organizing and managing such collaboration. Merely putting people with different backgrounds in a project, in a conference room, around a table, does not do the trick, alas. Extra efforts are required, for example, to promote curiosity, creativity, and collaboration (which we discussed in Chapters 16 and 17).

Moreover, it has become clear that the tech sector can be particularly unwelcoming (toxic, even) to diversity, for example, in terms of hiring people or promoting people. Notably, women and people of colour have been excluded and treated wrongly. Timnit Gebru, with an impressive track record in various technical roles in various firms, for example, was hired by Google and fulfilled a leading role in ethics of AI. Then, after co-authoring a scientific paper that discussed drawbacks of large language models, she was fired. Her firing sparked an industry-wide discussion of diversity. Let us hope that the industry can improve this. Key benefits of diverse project teams and diverse leadership are that they bring perspectives to the table that can benefit and advance the project. Diversity can prevent status quo serving tendencies that can stifle innovation.

Reflexivity

Lastly, we will discuss reflexivity, another key dimension of RI. Reflexivity refers to reflection on activities that you are actively involved in and on your own involvement in these activities. It involves becoming (more) aware of your own assumptions, values, and commitments and their effects on your own role in projects. Do you focus on delivering on *time*? Or on delivering *quality*? Are you concerned with short-term benefits? Or are you attuned to longer-term benefits? Are you oriented towards creating technology, as a means? Or towards the end of helping to solve a problem in society?

Being aware of your own assumptions, values, and commitments can help you to be more open towards other people's assumptions, values, and commitments. You can temporarily suspend your own thoughts and feelings and focus on what others say, what they think and feel. In that sense, reflexivity, and the ability and willingness to steer your own attention, is a precondition for all communication and collaboration. In a meeting, with team members, with customers (HCD), or with stakeholders (VSD), it will help enormously if you are aware of your own views and concerns and at the same time, can be open towards the other participants' views and concerns.

As a way to illustrate and promote reflexivity, let us look at the term *Responsible Innovation*. Can you see a tensions between these two words? On the one hand, RI requires a commitment to responsibility, which I associate with caution and care. On the other hand, RI requires a commitment to innovation, which I associate with risk-taking and boldness.

Responsibility and innovation are both needed in Responsible Innovation. For responsibility, you will need to cultivate virtues like justice, civility, care, and inclusion. You may want to imagine Rachel. She works as a physicist in a project that aims to accelerate the transition to sustainable energy. She stresses justice in the sense of making these solutions available and affordable for different groups of people, also those with less disposable income. She is keen to explore potential adverse effects of the innovation they work on, in order to create measures to mitigate these. Furthermore, she advocates organizing roundtable sessions to involve citizens, to learn about their needs, so they can take these into account. Additionally, for innovation, you will need to cultivate virtues like courage, curiosity, creativity, and collaboration. Here, you may want to imagine Iason. He works as a data scientist in a project that aims to improve the transparency of algorithms that are used by government agencies. He proposed the project; a brave move of him, out of his comfort zone. He is aware of the need to exercise self-control, for example, when he needs to refrain from adding features ('feature creep') and instead aims for simplicity and usability. He collaborates with people with diverse backgrounds, which requires curiosity and creativity. He asks open questions and endures many moments of relative uncertainty.

I can share two experiences that relate to reflexivity. At the start of my career, I used to organize focus groups to discuss and evaluate ideas for new telecommunication services. I remember one particular focus group. The client, of a telecom operator, had asked me to find out what people thought about several ideas for new services, including an idea to sell ringtones (popular, at that time), and, based on that, to provide advice for further development. Halfway the focus group, one participant turned to me and said: 'I do understand. They try to make teenagers addicted to this service. So they spend all their money. But you don't care of course. You work for them'. This was an awkward moment for me. As a parent, I did empathize with the concern he expressed. As a consultant, I felt caught red-handed. And as a researcher, I wanted to know more. It took me a couple of seconds to collect my thoughts and feelings. Then I chose to be open, I told a bit about my different roles: as parent, as consultant, and as researcher. Then I invited him to tell more about his concerns. Another, more recent, example is from an international project to promote RI in two Research and Technology Organizations. Near the end of the project, we concluded that reflexivity had been a critical success factor. You can take all sorts of actions, promote anticipation and responsiveness, bring together people with diverse backgrounds, collaborate with prospective users, but the effects of such actions depend on reflexivity. RI requires that you make room for reflexivity. Ask uneasy questions, share vulnerable experiences, deal with awkward moments, and tolerate uncertainty.

Reflexivity can be understood as a form of practical wisdom (*phronesis*), which deals with presence, awareness, and openness. Otto Scharmer (whom we already met in Chapter 13), talks about the need to cultivate an *open mind*, so you can be open towards other people and their ideas; an *open heart*, so you can be open towards other people's experiences and concerns; and an *open will*, so you can be moved to compassionate and practical action. This requires an awareness of your own thoughts, feelings, and concerns, and an ability and willingness to temporarily 'bracket' these, and instead be open towards other people's thoughts, feelings, and concerns.

In what ways do you currently involve societal actors or stakeholders, citizens or Civil Society Organizations in your project? If you imagine doing more of that, which benefits can you envision? And which challenges? In what ways are you organizing anticipation and responsiveness? And what about diversity, inclusion, and participation? How are these needed in your project? How much is currently there, and how may you improve that? Moreover, I would invite you to make (more) room for reflexivity. Make time for a walk, maybe in nature. Make a nice lunch and enjoy it. Maybe in the company of others. Take a nap. Take a shower. Now you have (more) room for reflexivity. How do you think and feel about your project and about your role in it? What would you like to change? What could be first steps to start that change?

Chapter 20

What does your next project look like?

I hope that you feel inspired and informed to integrate *doing ethics* in your projects and that you can better align your head, your heart, and your hands. Please feel free to use any of the approaches discussed and to combine elements from different approaches. You are invited to organize projects that are iterative, participatory, and inclusive, so that they can help to develop and deploy products, services, and systems that contribute to creating just societies in which people can flourish and live well together.

In closing, I would like to present a checklist. The checklist is a recurring trope in discussions of ethics in tech and innovation. Some would love to have a list, with items they can check and be done with. Others have objections to such a list, they

DOI: 10.1201/9781003088776-23

believe that ethics, especially ethics understood as a process of ethical reflection, inquiry, and deliberation, cannot be captured in such a list.

The checklist below is an attempt to combine these concerns; it lists activities that you can choose to organize, and that together can constitute process of *doing ethics*:

Reflect on the role of technology in your project. Avoid instrumental or deterministic views. How does the project aim to shape technology? And how can technology, in turn, shape what people can (not) do with it?

You can think of technologies as tools to enable people to develop human capabilities. What capabilities are relevant in your project? How do the outcomes of your project enable people to develop these capabilities?

How do you want your project to create value? In terms of well-being? In terms of economic development? You can combine these goals, for example, in sustainable development (people, planet, prosperity).

You can organize ethical reflection, inquiry, and deliberation; for example, by following the three-step process: identify moral issues, organize dialogues, and make tentative choices, which you can try-out and evaluate.

You can use consequentialism to identify and evaluate positive and negative outcomes and impacts; and to discuss and define system boundaries and externalities, what you will (not) take into account in your analysis.

You can use deontology to focus on human dignity and human autonomy; to identify duties and rights that are at stake and to deal with tensions between duties and rights, for example, related to different actors.

You can use relational ethics to understand people as social beings, connected to others and to nature; and to discuss the effects of technologies on interactions between people, and between people and the natural environment.

You can use virtue ethics to reflect on the ways in which your project's outcomes' can help or hinder people to live well together. In addition, it can help *you* to identify and cultivate relevant professional virtues.

From Human-Centred Design (HCD), you can borrow methods to organize an iterative and collaborative process, in which you can effectively combine innovation and ethical reflection, inquiry, and deliberation.

From Value Sensitive Design (VSD), you can borrow methods to involve relevant actors and stakeholders, so they can discuss values (matters they value), with the goal to integrate these into the innovation process.

From Responsible Innovation (RI), you can borrow methods to organize societal engagement, and to promote anticipation and responsiveness, participation, inclusion, and diversity, and make room for reflexivity (practical wisdom).

A recurring question is how you can promote HCD, VSD, or RI. You can allocate budgets to efforts like this. And there are methods available. But you also need to be able and willing to engage in such methods, for example, make room for iterations and collaboration, and make room for less-usual stakeholders at the table. It is worthwhile to identify and analyse enablers and barriers that are at play. Enablers you will want to utilize and strengthen. Barriers you will want to reduce and break down. Enablers and barriers can occur both in 'structure', for example, in work processes, organizational culture or shared norms, and in 'agency', for example, in people's practical skills, beliefs or dispositions. You can talk to diverse people, within your organization or in other organizations, for example, in suppliers or clients, or in academia or in government, in order to identify and analyse these, and to find ways to utilize enablers and reduce barriers.

I would like to invite you to share your experiences and findings with others and to explore what others have found.

Chapter 21

Exemplars

On the following pages are short portraits of people whom I find admirable. They embody, exemplify, or champion specific virtues. They can function as *exemplars* and inspire you to cultivate and express relevant virtues. You can watch online talks by them or listen to interviews with them. You can model their ways of thinking, feeling, and acting, and learn from them. In addition, you can try-out relevant virtues in your work and projects, and learn by doing. The aim is to learn to (better) align your thoughts, feelings, and actions, so that your virtues become habits.

Please note that I do *not* expect that you view all of these exemplars admirable in the same way that I do. You will

DOI: 10.1201/9781003088776-24

probably look differently at some of them. Maybe you even do not admire some of them. I do hope, nevertheless, that you can look at some of them and feel inspired. Please do feel free to add your own exemplars: people whom you admire and want to learn from.

Select a project that you work on. Imagine this project's impact in society and identify one or two virtues that are at stake, for example: justice or civility, etc. Pick one or two exemplars who embody these virtues. Watch an online talk or listen to an interview. Then try-out these virtues in your project: advocate for making the algorithm more fair, or propose to build features that facilitate civility, etc.

Greta Thunberg

In August 2018, after a series of heat waves and wildfires during Sweden's hottest summer since over 250 years, Greta Thunberg, then 15 years old, began her *Skolstrejk för klimatet, school strike for the climate*, sitting in front of the Swedish Parliament in Stockholm. She wanted the Swedish government to comply to the *Paris Agreement* and reduce carbon emissions. In the ensuing months and years, she has inspired millions of people worldwide to organize climate action and demand that governments and businesses act against the climate crisis. She has delivered a series of arresting speeches at key international conferences.

Thunberg embodies the virtue of *courage* because it takes lots of courage, strength, dedication, and perseverance to do what she does. She has a unique way of navigating between fear and hope. When she was 11 years old, she became depressed because she knew the facts about the climate crisis and she saw that most people in positions of power took very little action. She learned to transform her fear into positive action and inspiration for millions of people.

She also exemplifies the virtues of *justice* and *civility*. She broke away from social norms and demanded justice for future generations, insisting on their rights to inherit and inhabit a liveable planet. Moreover, she has been a significant force in mobilizing civil society organizations and to put the climate crisis on the agenda and at the centre of political and public debates.

Tristan Harris

In 2011, Tristan Harris started working at Google. There, he became increasingly concerned about the effects of digital and online products on people's daily lives and on society. The advertisement-based business models of companies like Google yield products that aim to maximize users' screen time, not their wellbeing. Apps are designed like slot machines; they give little rewards, and the resulting dopamine hits make you keep on using the apps. In 2015, he left Google to found *Time Well Spent*, which later became the *Center for Humane Technology*.

Harris exemplifies the virtue of *self-control*, which entails the setting of goals that contribute to wellbeing, and working conscientiously towards these goals. If we cultivate self-control, we can move away from behaving like we are tools for the benefit of the company that provides these technologies and move towards using these very same technologies as tools that promote human flourishing. Self-control is about using technology consciously and positively.

He also embodies the virtues of *civility* and *empowerment*. The *Center for Humane Technology* facilitates critical conversations about tech amongst developers, youth, parents and educators, and policy makers. These conversations aim to promote civility, to find ways to design and deploy technology in responsible ways. Moreover, they provide practical advice to empower people to change their relationship to tech, for example, by turning off notifications on their mobile phones, keeping phones out of the bedroom, and organizing 'device-free' diners.

Kate Raworth

Educated as an economist, Kate Raworth worked with micro-entrepreneurs in the villages of Zanzibar, co-authored the Human Development Report for the UN in New York, and then worked as Senior Researcher at Oxfam. She felt increasingly uncomfortable about the dominant paradigm of growth. She calls herself a 'renegade economist focused on exploring the economic mindset needed to address the 21st century's social and ecological challenges'. She published *Doughnut Economics: Seven ways to think like a 21st century economist.*

Raworth embodies the virtue of *justice*; notably, ecological and social justice. The *Doughnut*'s two concentric circles form the boundaries where we need to stay within in order to be able to ensure that no persons fall short on life's essentials (smaller circle), such as food, housing, health care, education, and political rights and to ensure that we do not overshoot the Earth's capacity to support life (larger circle), regarding a stable climate, fertile soils, abundant biodiversity, healthy freshwater cycles, and a protective ozone layer overhead.

She also exemplifies the virtues of *perspective* and *collaboration*. She wants to change how we view economics: to replace pictures of curves going up, often based on extraction and exploitation, with pictures of cyclic processes, based on regeneration and inclusion. Moreover, she launched the *Doughnut Economics Action Lab*, in 2020, to collaborate with changemakers worldwide. In 2020, she was appointed Professor of Practice at the Amsterdam University of Applied Sciences, to strengthen her collaboration with the City of Amsterdam.

Otto Scharmer

Otto Scharmer works as a senior lecturer at the MIT Sloan School of Management and is one of the co-founders of the *Presencing Institute*, an action research platform at the intersection of science, consciousness, and profound social and organizational change. He has introduced the *Theory U* framework and created a set of methodologies for systemic change and organizational learning.

Scharmer embodies the virtue of *empowerment*. His leadership style is serving, with a minimum of ego getting in the way. He has set-up multiple courses and training programs, both online and in person, via edX.org and the *Presencing Institute*. Through these courses and programs, he has inspired and empowered more than 250,000 people to bring about positive change. His methods and tools are applied by organizations in civil society, business, government, and United Nations agencies around the globe.

He also exemplifies the virtues of *perspective* and *care*. He provides new perspectives, to analyse how we currently organize our economies and societies, and to envision moving towards more sustainable and just societies. The 'U' shape visualizes the path from problem to solution: not directly from A to B, but via a curve, a journey. This journey demands of the people involved that they cultivate an open mind, an open heart, and an open will: curiosity, compassion, and courage. In his courses and webinars, you can experience how he creates a safe space for learning and how he carefully combines a vocabulary of business and technology with a vocabulary of personal change and organizational development.

Cathy O'Neil

After earning a PhD in math from Harvard, Cathy O'Neil was a postdoc at MIT, a professor at Barnard College, and a *quant* for the hedge fund D.E. Shaw during the credit crisis. She left finance in 2011 and started working as a data scientist in the New York start-up scene. In 2016, she published *Weapons of Math Destruction: How Big Data Increases Inequality and Threatens Democracy*. It reads like a catalogue of everything that can go wrong with algorithms and it is a must-read for anyone who is interested in, or works in, data science.

O'Neil embodies the virtues of *justice* and *honesty*, which go hand in hand for her. She knows the limits of what algorithms can do and cannot do. She debunks excessive promises that often surround algorithms. 'Promising efficiency and fairness, [algorithms] distort higher education, drive up debt, spur mass incarceration, pummel the poor at nearly every juncture, and undermine democracy' (*Weapons of Math Destruction*, p. 199).

She also champions the virtue of *care*. She warns us for algorithms that propagate and exacerbate injustices, for example, through using invalid proxies or deficient feedback loops. Or by using variables that can be quantified easily or data that can be collected easily. These variables and data do not necessarily represent what is actually important. Moreover, she provides practical advice for designing more just algorithms.

Douglas Rushkoff

A media theorist, author, lecturer, graphic novelist, and documentary maker, Douglas Rushkoff has been writing about online media since the 1990s. In a more recent conference panel, Rushkoff countered an interlocutor's transhumanist ideas and said that he values humanity, including our human quirks. The panellist remarked 'You're only saying that because you are human', to which Rushkoff declared: 'Yes, I am on *Team Human*'. This became the title of his 2019 book *Team Human*. In it, he brings together key ideas that he developed over the years, in a dozen bestselling books on media, technology, and culture. He also hosts the *Team Human* podcast and community events, which enable people 'to find the others'.

'Our technologies, markets, and cultural institutions – once forces for human connection and expression – now isolate and repress us. It's time to remake society together, not as individual players but as the team we actually are' is on the book's cover; his manifesto in a nutshell.

Rushkoff embodies the virtues of *flexibility* and *creativity*. He has been a pioneer of cyberculture, but when the early, non-profit internet was overtaken by venture capitalism and corporate interests, he became a critic of the digital and online utopia. He also exemplifies the virtues of *civility* and *care*. He fosters conversations, sometimes provokingly, but always carefully, about what we would want to aspire towards. He proposes a humanistic approach to technology, to infuse technology with what makes us uniquely human.

Marleen Stikker

In 1993, Marleen Stikker brought the Internet to the Netherlands; she founded *De Digitale Stad (The Digital City*; nominated, in 2020, for the *Unesco Memory of the World* list), in Amsterdam. This involved installing a series of internet terminals in public buildings, on which citizens could create virtual homes and communicate with others. Several years later, she co-founded a centre for research and innovation in the public domain, notably in culture, health care, and education. First named *Society for Old and New Media*, it is now known as *Waag, a Future Lab for technology and society*, named after the 15th-century weighing house it is located in.

Stikker embodies the virtue of *collaboration*, notably between diverse organizations and between people with diverse disciplinary backgrounds. This is key when it comes to questioning and understanding our current challenges from different perspectives and when it comes to mobilizing and combining different fields of expertise to explore and create solutions.

Furthermore, she exemplifies the virtues of *justice* and *civility*. In her book, *The Internet is broken; but we can repair it* (in Dutch), she reflects on the Internet's original spirit of creativity and conviviality and on the ways in which corporations introduced a focus on commerce, profit, and winner-takes-all strategies. Moreover, she advocates an approach to using the internet in ways that prioritize public values, as an alternative to business- or state-dominated approaches.

Jaron Lanier

In the 1980s, Jaron Lanier pioneered Virtual Reality and founded VPL Research to produce and sell equipment for VR. From 2009, he has worked at Microsoft Research as an Interdisciplinary Scientist. He wrote several books in which he critically engages with digital and online technologies, like: *You Are Not A Gadget, Who Owns the Future? Dawn of the New Everything*, and *Ten Arguments for Deleting Your Social Media Accounts Right Now* (2018). He is also a musician.

Lanier embodies the virtue of *perspective*. He has a humanistic vision on technology, and his perspective ranges from attention for small things, like a user interface detail that supports or stifles communication, so that it results in, for example, a civil conversation or in toxic mob behaviour, to attention to large things, like the ways in which digital and online technologies are designed in ways to grab and manipulate people's attention, thereby damaging the fabric of society.

In addition, he exemplifies *justice* and *flexibility*. Lanier worries about the inequalities that many digital and online services propagate and exacerbate. Silicon Valley's belief that 'information wants to be free' has let to one-dimensional, ad-driven business models, which amass huge profits for a very small number of people. He flexibly moves back and forth between, on the one hand, action and entrepreneurship, and, on the other hand, reflection and academic work. He is simultaneously a friend and a critic of Silicon Valley.

Sherry Turkle

For almost 40 years, Sherry Turkle, professor of the Social Studies of Science and Technology at MIT, has studied how people relate to computers. She wrote an impressive series of best-selling books, among which: *The Second Self: Computers and the Human Spirit* (1984), *Life on the Screen: Identity in the Age of the Internet* (1995), *Alone Together: Why We Expect More from Technology and Less from Each Other* (2011), *Reclaiming Conversation: The Power of Talk in a Digital Age* (2015), and *The Empathy Diaries: A Memoir* (2021).

In *Reclaiming Conversation*, she discusses how our obsession with being 'always on' made us forget the benefits of solitude and the value of connecting to others. This has corroded our skills for conversation. Critically, we need to ameliorate this situation; this can be done by cultivating skills for having and hosting conversations.

Turkle exemplifies the virtue of *civility*: she is committed to promoting collective deliberation about the good life and to foster collaboration towards in order to live well together. In addition, she champions the virtues of *empathy* and *care*. With her analyses and proposals, she urges us to cultivate empathy, to cultivate the skills of conversation, and to cultivate care between different people and groups of people. She is not against technology, but pro conversation.

Yuval Noah Harari

A professor in the Department of History at the Hebrew University of Jerusalem, Yuval Noah Harari is best known for his books. In *Sapiens: A Brief History of Humankind*, he writes about how we have created subjective realities: we believe in ideas and thereby make them real. People built pyramids and empires, invented money, and the rule of law. In *Homo Deus: A Brief History of Tomorrow*, he writes about the merging of infotech and biotech: integrating computers into our bodies and delegating tasks to computers.

Harari embodies the virtue of *humility*. He acknowledges the limits of our knowledge and abilities. We, humans, are remarkably similar to non-human animals. In addition, he champions the virtue of *empathy*. He follows a vegan diet to reduce the mass scale torturing and killing of cows, pigs, and chickens. He also uses our cruel treatment of animals as a cautionary tale for how cruelly some future artificial, superior lifeform may, one day, treat us.

Moreover, he exemplifies the virtue of *perspective*. In his books, he fashions novel perspectives on the human condition. Interestingly, he practices Vipassana meditation to cope with the deluge of thoughts and to have a clearer perspective on reality. Another example is the way he talks about his experience of being gay. He learned to distinguish between some people's belief that 'homosexuality is wrong' and the biological fact that 'homosexuality is present in many species', and to dismiss the former and embrace the latter.

Mariana Mazzucato

A professor in the Economics of Innovation and Public Value at University College London and founding director of the UCL Institute for Innovation and Public Purpose, Mariana Mazzucato is the author of *The entrepreneurial state: Debunking public vs. private sector myths*; *The value of everything: Making and taking in the global economy*; and *Mission economy: A moonshot guide to changing capitalism*. These books can be read as a trilogy, in which she advocates organizing better collaborations between government, industry, and societal actors. More specifically, she promotes *mission-oriented innovation*: to create innovations that contribute to more sustainable economies and just societies; to better align our innovation efforts with the UN Sustainable Development Goals.

Mazzucato exemplifies the virtues of *anticipation* and *responsiveness*. She has supported numerous organizations in looking ahead and moving towards responsive action, notably in her advocacy for *mission-oriented innovation* and in multiple advisory roles, for example, for the European Commission, in helping them to shape their research and innovation programmes.

She also champions the virtues of *honesty* and *courage*. She is honest and, indeed, very knowledgeable, about the ways in which finance works; how financialization, unchecked, can lead to exploitation and injustices. In addition, she enables people to envision alternative ways of organizing economies and societies; this requires courage and hope, to go against the stream.

John C. Havens

John C. Havens is a multi-talented person. He has been a professional actor, business consultant and speaker, journalist and author, and founder of several initiatives to promote wellbeing. He is also executive director of the *IEEE Global Initiative on Ethics of Autonomous and Intelligent Systems*, which delivered recommendations for *Ethically Aligned Design*, co-authored by some 600 experts from around the globe.

He embodies the virtues of *empathy* and *compassion*. He expresses these virtues both on the international level, for example, in his concerns for global issues and in his efforts to help steer innovation towards promoting people's wellbeing and on the practical level of interacting with others, for example, in his welcoming way to invite and include people with diverse backgrounds in *IEEE* working groups. He would probably refer to these virtues as *kindness*. This is the focus of his current interest: to study and promote self-directed and other-directed kindness.

In addition, Havens exemplifies the virtues of *collaboration,* *diversity,* and *inclusion.* He is keen to promote collaboration, for example, in his work for *IEEE,* which involves collaboration between people from academia, industry, and policy making. He also promotes diversity and inclusion, for example, by inviting people with different qualities and backgrounds and fostering mutual learning. Fun fact: he also has brought playfulness to multiple conferences or meetings, playing blues tunes on his harmonica.

Safiya U. Noble

An associate professor of Gender Studies and African American Studies at the University of California, Los Angeles (UCLA), where she serves as co-founder and co-director of the UCLA Center for Critical Internet Inquiry, Safiya U. Noble is known for her book *Algorithms of Oppression: How Search Engines Reinforce Racism.* It contains numerous examples of shocking search results, ranging from Google's image recognition that tagged a picture of two black teenagers with 'gorillas' to Google Maps that pointed to the White House, during Barack Obama's presidency, when looking for 'N*gga House', to pictures of white women that pop up when searching for 'professional hairstyles for work', and pictures of black women when searching for 'unprofessional hairstyles for work'.

In her work, Noble champions the virtues of *diversity* and *inclusion.* Notably, she critiques the application of biased algorithms, which produce racist and sexist outcomes—algorithms that marginalize the experiences of minority groups and deteriorate their opportunities.

Her advocacy for diversity and inclusion is aligned with her championing of *honesty* and *justice.* She unveils how search engines, which are often regarded as 'objective', are, in fact, very often skewed, biased, prejudiced—or, in short: *unjust.* Moreover, with a background in Library and Information Science, she is well aware of the importance of accessible and truthful information and the critical role of information as a condition for democracy and justice.

John Tasioulas

After studying philosophy and law at the University of Melbourne, John Tasioulas received a Rhodes Scholarship, and moved to Oxford, to become an expert in moral, legal, and political philosophy. He is Professor of Ethics and Philosophy of Law at the University of Oxford and the inaugural Director of their Institute for Ethics in AI. He is known for his work on the philosophy of human rights; he has proposed to understand these as moral rights, with foundations both in human dignity, which all people share, and in obligations to meet diverse human needs.

Tasioulas exemplifies the virtue of *collaboration*. Throughout his career, he has promoted collaboration between experts from different fields. In a recent essay, for the *Ada Lovelace Institute*, he advocated for involvement of the arts and humanities in our thinking about AI. In it, he outlines three tenets for a humanistic approach to the ethics of AI: *pluralism*, to embrace a plurality of values in our understanding and promoting of human wellbeing and morality; *procedures*, to care about 'not just the value of the outcomes that AI can deliver, but the processes through which it does so'; and *participation*, to enable diverse stakeholders to participate in decision making related to the design and deployment of AI.

He also champions the virtues of *diversity* and *justice*. The aim of such a humanistic approach, and of his work more broadly, is to promote diversity and justice.

Timnit Gebru

In December 2020, Timnit Gebru was fired by Google, where she worked as co-lead of the Ethical Artificial Intelligence Team. Her firing was triggered by a paper they wrote about language models, such as GPT-3; they compared these to parrots, in that they repeat combinations of words from the training data, which makes them susceptible to bias in these training data. The incident is one of many examples of the inabilities of tech companies to deal with critique of their 'normal' ways of working; their inabilities to uphold justice.

Gebru champions the virtue of *justice*. She has helped pioneer research into facial recognition bias against people of colour.

Inspired by, for example, the ProPublica research into COMPAS (the recidivism-risk algorithm with racial bias), she has been at the forefront of critiquing such software.

In 2021, she launched the *Distributed Artificial Intelligence Research Institute* (DAIR), which aims to not only document such biases and harms, but also to develop applications that can instead have positive impacts on disadvantaged people. DAIR promotes 'Community, not exploitation', 'Comprehensive, principled processes', and 'Proactive, pragmatic research' and aims to be a place for 'Healthy, thriving researchers'. She fosters collaboration between organizations with similar goals, for example, *Data & Society*, *Algorithmic Justice League*, and *Data for Black Lives*. In her work, she exemplifies also the virtues of *diversity* and *empowerment*.

Edward Snowden

In 2013, Edward Snowden, while working for US Central Intelligence Agency (CIA), copied and leaked classified documents about the National Security Agency, notably about global surveillance programs run by the NSA and *Five Eyes*, an intelligence alliance between the US, the UK, Australia, Canada, and New Zealand. The leaked documents revealed that these countries had been spying on one another's citizens and had shared the collected intelligence, in order to circumvent restrictive domestic regulations on the surveillance of citizens.

Some call him a patriot and a hero. Others call him a traitor and a danger to government. For me, he embodies the virtue of *care*. It is exactly because he cares about security and privacy that he became a whistle blower. In a 2014 TED Talk, he argued that 'We don't have to give up our privacy to have good government. We don't have to give up our liberty to have security'.

He also exemplifies the virtues of *civility* and *courage*. In this TED Talk, Snowden explained that he sees himself primarily as a citizen: a citizen who is concerned with privacy and transparency. He saw the dangers of mass surveillance and acted out of courage, for the common good, and has paid a high price; he had to flee prosecution in the US and currently lives in exile in Russia.

Notes and further reading

Chapter 1. A humanistic approach

- High-Level Expert Group on Artificial Intelligence (2019): *Ethics guidelines for trustworthy AI*. Written for the European Commission, this document provides an introduction into a human-centric approach to the design and deployment of AI systems; it discusses ethical principles like respect for human autonomy, prevention of harm, fairness, and explicability.
- IEEE (2019): *Ethically aligned design*: *A vision for prioritizing human well-being with autonomous and intelligent systems* (first edition). Useful resource, co-authored by hundreds of experts from industry and academia; it includes a series of principles, related to human rights, wellbeing, data agency, effectiveness, transparency, and accountability.
- Wendell Wallach and Colin Allen (2009): *Moral machines*: *Teaching robots right from wrong*. They discuss the need for ethical sensitivity in the design and deployment of machines; it explores the tension between autonomous systems and human autonomy, and top-down and bottom-up ways to implement machine ethics and human oversight.
- Donella Meadows (2008): *Thinking in systems*: *A primer*. An excellent and accessible introduction to systems thinking, by lead-author of the bestselling *Limits to growth*, which helped to put the need for sustainable development on the agenda. This book may very well change your outlook at the world; it will help you to see both the forest and the trees.

- David Hildebrand (2008): *Dewey: A beginner's guide.* Excellent and relatively accessible introduction to the ideas of John Dewey. The book discusses his ideas on experience, inquiry, morality, politics, education, and aesthetics. There is a growing interest in philosophical pragmatism, for example, in organization studies and design studies.

Chapter 2. What do we mean with *ethics*?

- Bruno Latour (1987): *Science in action: How to follow scientists and engineers through society.* An accessible introduction to the empirical study of science and technology, 'in action', 'in the making'. More recently, Latour wrote about the climate crisis and policies to counter it: *Facing Gaia* (2018) and *Down to Earth* (2019).
- Bruno Latour (1980): ***Three little dinosaurs or a sociologist's nightmare.*** You can find the short essay here: http://www.bruno-latour.fr/sites/default/files/07-DINOSAURES-GB.pdf
- Bent Flyvbjerg (2001): ***Making social science matter: Why social inquiry fails and how it can succeed again.*** A more advanced or scholarly book about an approach to combine social science and natural science in a manner that does justice to their respective qualities; where natural science can talk about facts, and social science can talk about values.
- Marc Steen (2013): **Co-design as a process of joint inquiry and imagination,** *Design Issues*, 29 (2), 16–28 (open access). Here, I apply Dewey's ideas on joint inquiry to the context of design and innovation: in the TA2 project. I also discuss induction, deduction, and abduction.

Chapter 3. Is technology a neutral tool?

- Langdon Winner (1993): **Upon opening the black box and finding it empty:** Social constructivism and the philosophy of technology, *Science, Technology, & Human Values*, 18 (3), 362–378. In this article, Winner discusses the low-hanging bridges. He published multiple books on the interplay between politics and technology, for example, *The whale and the reactor: A search for limits in an age of high technology* (1988; second edition 2020).

- Marc Steen (2015): **Upon opening the black box and finding it full**: Exploring the ethics in design practices, *Science, Technology, and Human Values*, 40 (3), 389–420. In this article (the title of which alludes to Winner's 1993 article), I explore the ethics that are inherent in the practices of design, technology development, and innovation.
- Nelly Oudshoorn and Trevor Pinch, editors (2005): **How users matter**: *The co-construction of users and technology*. Relatively accessible book about the active and creative roles of users in shaping technologies, during design, adoption, and usage.
- Peter-Paul Verbeek (2011): **Moralizing technology**: *Understanding and designing the morality of things*. A more scholarly book about the ethical issues that are at play in designing and interacting with technologies, with practical examples.
- Jaron Lanier (2019): **Ten arguments for deleting your social media accounts right now.** Lanier argues that we need to move away from the grip of social media and their toxic business models, if we want to save our societies and personal lives.
- Douglas Rushkoff (2019): **Team human.** Rushkoff argues that 'it's time to remake society together, not as individual players but as the team we actually are' (from the book's cover). His tone can be sharp in places, but his overall message is one of compassion.
- Sherry Turkle (2015): **Reclaiming conversation**: *The power of talk in a digital age.* Turkle has studied people interacting with technologies for 40 years. In this book, she discusses the value of conversations. She is not against technology, but pro conversation.
- Meredith Broussard (2018): **Artificial unintelligence**: *How computers misunderstand the world.* Award-winning writing style; 'a guide to understanding the inner workings and outer limits of technology'. I especially enjoyed the lively storytelling.

- Safiya Umoja Noble (2018): *Algorithms of oppression*: *How search engines reinforce racism*. Noble discusses how supposedly neutral technologies can actually produce discriminatory and harmful outcomes, with many very disturbing examples.
- Brett Frischmann and Evan Selinger (2018): *Re-engineering humanity*. More scholarly, but rather accessible. An image that stuck with me: the risk of people behaving like machines (an inversion of the more common image of making machines behave like people).
- Catherine D'Ignazio and Lauren F. Klein (2020): *Data feminism*. 'A new way of thinking about data science and data ethics that is informed by the ideas of intersectional feminism'. Available online, open access: https://data-feminism. mitpress.mit.edu/
- Virginia Eubanks (201): *Automating inequality*: *How high-tech tools profile, police, and punish the poor*. Eubanks 'systematically investigates the impacts of data mining, policy algorithms, and predictive risk models on economic inequality and democracy in [the US]'.

Chapter 4. Value, wellbeing, and economics

- Martin Seligman (2011): *Flourish*: *A new understanding of happiness and wellbeing*. Accessible introduction to positive psychology by one of its pioneers. Notably, this book discusses a range of practical examples and recommendations.
- The New Economics Foundation 'works with people igniting change from below and combines this with rigorous research to fight for change at the top'. They published many reports, for example, on wellbeing and on technology: https://neweconomics.org/section/publications
- Ingrid Robeyns (2017): *Wellbeing, freedom and social justice*: *The capability approach re-examined* (available online, open access). Chapters 1–2 can function as a relatively accessible introduction to core ideas; Chapters 3–5 provide more advanced discussions.
- Ilse Oosterlaken (2009): **Design for development**: A capability approach, *Design Issues*, 25 (4), 91–102 (open access).

Great introduction to the CA in the domain of technology and innovation, by a pioneer. Available online: https://doi. org/10.1162/desi.2009.25.4.91

- Ilse Oosterlaken and Jeroen van den Hoven (2012): *The capability approach, technology and design* (available online, open access). More scholarly book, if you want to learn more about applying the capability approach in technology and innovation.
- Marc Steen (2016): **Organizing design-for-wellbeing projects**: Using the capability approach, *Design Issues*, 32 (4), 4–15. Here, I discuss, both theoretically and practically, how one can apply the CA in the field of technology and innovation.
- For more about the Capability Approach, you can turn to Martha Nussbaum or Amartya Sen; for example, Sen's books *Development as freedom* (1999) or *The idea of justice* (2009; with lots of advanced economics, but also more accessible sections), or Nussbaum's *Creating capabilities* (2011; with, for example, a list of 'central capabilities', on pages 33–34).
- Kate Raworth (2018): *Doughnut economics*: *Seven ways to think like a 21st-century economist.* A very accessible and inspiring book, with lots of images, especially if you are a visual thinker; it can enable you to view the world differently. The quote of Kuznets is on page 40 and is from Kuznets (1962): 'How to judge quality', in Croly (ed.), *The New Republic*, 147 (16), 29. The quote at the end of the chapter is from page 286.
- Mariana Mazzucato: *The entrepreneurial state*: *Debunking public vs. private sector myths* (2013); *The value of everything*: *Making and taking in the global economy* (2018); and *Mission economy*: *A moonshot guide to changing capitalism* (2021). The quote at the end of the chapter is from *Mission economy*, page 212.

Chapter 5. The problem with the 'Trolley Problem'

- Philippa Foot (1967): **The problem of abortion and the doctrine of the double effect**, *Oxford Review*, 5. Foot describes the original version of 'the trolley problem'; available at: https://philpapers.org/archive/FOOTPO-2.pdf

- Various different takes on the Trolley Problem can be found online, for example, here: https://en.wikipedia.org/wiki/Trolley_problem
- Edmond Awad and others (2018): **The moral machine experiment**, *Nature*, 563, 59–64 (https://www.nature.com/articles/s41586-018-0637-6). An extended version of this article, with additional information and analysis, is available at: https://ore.exeter.ac.uk/repository/handle/10871/39187

Chapter 6. Privacy is about more than 'privacy'

- Carissa Véliz (2020). *Privacy is power: Why and how you should take back control of your data.* An excellent discussion of different elements of privacy. Critically, Véliz connects privacy to other values that are required for human flourishing. The quote about transforming power is from page 51.
- About the *SyRI* case: https://www.rechtspraak.nl/Organisatie-en-contact/Organisatie/Rechtbanken/Rechtbank-Den-Haag/Nieuws/Paginas/SyRI-legislation-in-breach-of-European-Convention-on-Human-Rights.aspx
- Jaap-Henk Hoepman (2021): *Privacy is hard and seven other myths: Achieving privacy through careful design.* Hoepman discusses a range of effective solutions and strategies to protect and promote privacy, with lots of practical examples.

Chapter 7. What is your responsibility?

- Herman Tavani (2013): *Ethics and technology: Controversies, questions, and strategies for ethical computing.* An accessible text book, written for students, with many practical examples and pedagogical materials. I borrowed the 'discussion stoppers' from pages 42–51.
- The image of the rock climber and my discussion of responsibility as a combination of knowledge and agency is inspired by Mark Alfano's 2016 book *Moral psychology*. On pages

58–60 Alfano describes this combination as *chimneying*: 'a method for climbing a narrow vertical tunnel by pushing with one's right arm and leg on one side of the tunnel, while pushing with one's left arm and leg on the other side of the tunnel; to chimney is to exploit one constraint to put oneself in a position to exploit another constraint, which then allows one to exploit the first constraint again, and so on. In this case, the constraints aren't rock faces but the conceptual limitations imposed by knowledge and control [agency]' (page 59).

Chapter 9. Consequences and outcomes

- Ibo Van de Poel and Lambèr Royakkers (2011): *Ethics, technology and engineering: An introduction*. An excellent text book, widely used. I borrowed their example of the *Ford Pinto* (pages 67–69), which they discuss from different ethical perspectives. They discuss roughly the same four ethical perspectives as I do: utilitarianism, Kantian theory, care ethics (which I discuss in terms of *relational ethics*), and virtue ethics.
- For more about QALY: https://en.wikipedia.org/wiki/Quality-adjusted_life_year
- Peter Singer (2009, 2019): *The life you can safe* (10th anniversary edition) (available online: https://www.thelife youcansave.org). The book opens with the story of the child in the pond.
- Katarzyna de Lazari-Radek and Peter Singer (2017): *Utilitarianism: A very short introduction*. I find these 'very short introductions' useful, both accessible and rigorous. If you want to dig deeper into utilitarianism, you can find primary texts of, for example, Bentham or Mill, available at: https://www.gutenberg.org

Chapter 11. Duties and rights

- The example of giving money to the person who begs for money comes from a lecture of Marianne Talbot; from her podcast series *'A Romp Through the History of Philosophy'*. Talbot is Director of Studies in Philosophy at University of Oxford.

- Roger Scruton: *Kant*: *A very short introduction*. I found this 'very short introduction' less accessible. Probably Kant is relatively difficult. If you are interested, you may want to go directly to the chapter on the categorical imperative.
- *ACM Code of Ethics and Professional Conduct* (https:// www.acm.org/code-of-ethics). Another example, the *IEEE Code of Ethics* (https://www.ieee.org/content/dam/ieee-org/ ieee/web/org/about/corporate/ieee-code-of-ethics.pdf).
- John Tasioulas (2021): *The role of the arts and humanities in thinking about artificial intelligence (AI)* (blog post: https://www.adalovelaceinstitute.org/blog/role-arts-humanities-thinking-artificial-intelligence-ai/). Also worth reading, on the same site, Shannon Vallor (2021): *Mobilising the intellectual resources of the arts and humanities.* (https://www.adalovelaceinstitute.org/blog/ mobilising-intellectual-resources-arts-humanities/).
- Andrew Clapham (2015): *Human rights*: *A very short introduction* (2nd edition). Nice overview of the history of the development and implementation of human rights. It has, for example, a chapter on balancing rights for free speech with rights for privacy.
- Online resources of the United Nations' *Office of the High Commissioner for Human Rights* (https://www. ohchr.org/en/topic/business-and-human-rights) and of the *Organisation for Economic Co-operation and Development* (http://www.oecd.org/industry/inv/responsible-business-conduct-and-human-rights.htm).
- **Google apologises for Photos app's racist blunder** (https:// www.bbc.com/news/technology-33347866). One of the critiques of Google's 'fix' (https://www.wired.com/story/ when-it-comes-to-gorillas-google-photos-remains-blind/).
- Cathy O'Neil (2016): *Weapons of math destruction*: *How big data increases inequality and threatens democracy*. A must-read for anyone who works in data science or in an organization that uses data (that is: any organization).
- ProPublica (2016): *Machine bias* (https://www.propublica. org/article/machine-bias-risk-assessments-in-criminal-sentencing). One of the first articles on COMPAS, by investigative journalists of ProPublica.

- Mireille Hildebrandt (2020): *Law for computer scientists and other folk*. Nice introduction, with little legal jargon; covers topics like privacy, data protection, cybercrime, copyright, and legal personhood. Available online, open access: https://lawforcomputerscientists.pubpub.org
- Sheila Jasanoff (2016): *The ethics of invention*: Technology and the human future. A pioneer of *Science and Technology Studies*, Jasanoff discusses a range of cases, like the 1984 Bhopal gas disaster, and analyses their causes, dynamics, and outcomes, with particular attention to the role of law (which, sadly, does not always work to promote justice).

Chapter 13. Relationships and care

- Sabelo Mhlambi (2020): *From rationality to relationality*: Ubuntu as an ethical and human rights framework for Artificial Intelligence governance. Mhlambi is one of the pioneers of the application of Ubuntu philosophy to the design and application of AI.
- Tyson Yunkaporta (2020): *Sand talk*: How indigenous thinking can save the world. Challenges many taken-for-granted assumptions. I found Yunkaporta's storytelling engaging and interesting; he also reflects on the process of creating knowledge and of writing.
- Robin Wall Kimmerer (2013): *Braiding sweat grass*: Indigenous wisdom, scientific knowledge and the teaching of plants. Fine introduction to indigenous wisdom from North America. Kimmerer relays many stories about plants; but these stories are about much more.
- *Constitution of the Republic of Ecuador* (2008). The first national constitution that explicitly mentions rights of nature, for example, in article 71: 'Nature, or Pacha Mama, where life is reproduced and occurs, has the right to integral respect for its existence and for the maintenance and regeneration of its life cycles, structure, functions and evolutionary processes'.
- Daniel K. Gardner (2014): *Confucianism*: A very short introduction. Interestingly, Gartner draws parallels between different Confucian thinkers and different Western thinkers;

some focused on strengthening positive impulses, whereas others focused on correcting negative impulses.

- Pak-Hang Wong and Tom Xiaowei Wang, editors (2021): *Harmonious technology*: *A Confucian ethics of technology.* Edited book, with chapters that apply ideas from Confucian traditions on issues in technology, like personhood, technological mediation, and AI.
- Soraj Hongladarom (2020): *The ethics of AI and robotics*: *A Buddhist viewpoint.* Presents a perspective on the design and application of AI, based on insights from Buddhist teachings; it aims to help create ethical guidelines for AI that are practical and cross-cultural.
- Jason Edward Lewis, editor (2020): *Indigenous protocol and artificial intelligence position paper.* One of the first efforts to apply different traditions of indigenous knowledge to the discussion, design, and application of AI systems.
- Multiple scholars point at Francis Bacon as a starting point for the European Enlightenment in science and technology, e.g., Dutch professor emeritus Hans Achterhuis.
- The quote 'Nature must be taken by the forelock', by Francis Bacon, is referred to by, for example, Carolyn Merchant (1980): *The Death of Nature: women, ecology and the scientific revolution,* page 170. Apparently, 'Knowledge is power', while typically attributed to Francis Bacon, was coined by Thomas Hobbes (https://en.wikipedia.org/wiki/Scientia_potentia_est).
- Otto Scharmer and Katrin Kaufer (2013): *Leading from the emerging future: From ego-system to eco-system economies.* The three divides are on pages 4–5 (the interpretations of which are partly mine). An excerpt is available online, https://www.bkconnection.com/static/Leading_From_the_Emerging_Future_EXCERPT.pdf
- For an example of applying Levinas' ideas on the relation between self and other on technology development, you may want to have a look at: Marc Steen (2012): **Human-centred design as a fragile encounter**, *Design Issues*, 28 (1), 72–80. Dutch professor Vincent Blok, for example, also applied Levinas' ideas to (responsible) innovation.

- Michael Sandel (2009): *Justice: What's the right thing to do?* Accessible introduction to moral philosophy, combined with political philosophy of, for example, John Rawls. You can find Sandel's lecture series and other material online at https://justiceharvard.org/
- Frans de Waal published widely on animals, notably primates, also in relation to people and morality, for example: *The bonobo and the atheist: In search of humanism among the primates* (2014); or *Are we smart enough to know how smart animals are?* (2017).
- Rutger Bregman (2020): *Humankind: A hopeful history.* Bregman argues that most people are likely to cooperate. He debunks several psychology experiments that claimed the opposite. He discusses how news works; it focuses on the exceptions, on criminal, and cruel incidents.
- Virginia Held (2006): *The ethics of care: Personal, political, and global.* A very accessible and inspiring introduction to this ethical perspective. The quote is from page 64.
- Nicholas Carr (2010): *The shallows: How the internet is changing the way we think, read and remember.* Carr discusses various nefarious effects of using online devices and services (there is an updated 10th anniversary edition).
- Kate Darling (2021): *The new breed: What our history with animals reveals about our future with robots.* Nice discussion of how we collaborated with animals, like horses, and how that can help us to think about collaborating with robots. I find this view much more feasible than a view in which (humanoid, 'white and shiny') robots are required to mimic people.

Chapter 15. Virtues and flourishing

- Aristotle: *Nicomachean ethics.* Rather accessible, if you make time for it, and provided that you find a good translation, preferably one with chapter introductions and explanatory notes.
- John Thackara (2006): *In the bubble: Designing in a complex world.* A critical study of technologies' effects on society and wellbeing. The quote is from page 189.

- Shannon Vallor (2016): *Technology and the virtues: A philosophical guide to a future worth wanting.* Highly recommended reading. The book is both theoretically sound and relatively accessible. Vallor discusses virtue traditions in Europe (Aristotle), in Confucianism, and in Buddhism with equal depth and applies virtue ethics to four domains of practice: social media, surveillance, 'robots at war and at home', and human enhancement. Vallor discusses the following *technomoral* virtues: honesty, self-control, humility, justice, courage, empathy, care, civility, flexibility, perspective, magnanimity, and technomoral wisdom. The definition of self-control is from page 124, of empathy from page 133, and of civility from page 141.
- The *Facebook Home "Dinner"* video can be found here: https://youtu.be/yF3Nk4YlU_Y. Shannon Vallor refers to a 2013 article by Evan Selinger in *Wired*, in which he also discusses this video: 'Facebook Home Propaganda Makes Selfishness Contagious'. A more recent video, *Welcome to Facebook Horizon*, can be discussed with similar themes and concerns: https://youtu.be/Vj2q6KhNA_I
- Marc Steen (2013): **Virtues in participatory design:** Cooperation, curiosity, creativity, empowerment and reflexivity, *Science and Engineering Ethics*, 19 (3), 945–962. An early attempt of mine to apply virtue ethics to the domain of technology development.
- Marc Steen, Martin Sand, and Ibo Van de Poel (2021): **Virtue ethics for Responsible Innovation**, *Business and Professional Ethics Journal*, 40 (2), 243–268 (open access). We explore virtues that people in Responsible Innovation would need to cultivate, and present practical wisdom (*phronesis*, reflexivity) as key virtue.

Chapter 16. Methods to 'do ethics' in your project

- Marc Steen, Martijn Neef, and Tamar Schaap (2021): **A method for Rapid Ethical Deliberation in research and innovation projects**, *International Journal of Technoethics*,

12 (2), 72–85 (open access). We present a workshop format for Rapid Ethical Deliberation, in which participants can engage with four different ethical perspectives.

- Mark Alfano (2016): *Moral psychology: An introduction.* My discussion of patiency, agency, sociality, reflexivity, and temporality in relation to the four different ethical perspectives (which Alfano refers to as: utilitarianism, Kantian, care ethics, and virtue ethics) is based on Alfano's discussion on pages 1–21.

Chapter 17. Human-Centred Design

- Marc Steen (2013): **Co-design as a process of joint inquiry and imagination,** *Design Issues,* 29 (2), 16–28 (open access). A case study of managing an iterative process of problem-setting and solution-finding, in the TA2 project, informed by Dewey's pragmatism.
- Marc Steen, Jan Buijs and Doug Williams (2014): **The role of scenarios and demonstrators in promoting shared understanding in innovation projects,** *International Journal of Innovation and Technology Management,* 11 (1) (https://www.worldscientific.com/doi/epdf/10.1142/S021 987701440001X); on the iterative process in the TA2 project.
- ISO (2019): *Human-centred design for interactive systems,* ISO 9241-210:2019. This standard 'provides requirements and recommendations for human-centred design principles and activities throughout the life cycle of computer-based interactive systems'; see https://www.iso.org/standard/77520.html
- Marc Steen (2011): **Tensions in human-centred design,** *CoDesign,* 7 (1), 45–60. An overview of methods: participatory design, ethnography, lead user approach, contextual design, codesign, and empathic design. Presented in a two-dimensional grid, to understand them as connecting project team members and users (left-right) and connecting concerns for understanding current situations ('research') and envisioning future situations ('design') (top-bottom).

Chapter 18. Value Sensitive Design

- Batya Friedman and David Hendry (2019): *Value sensitive design*: *Shaping technology with moral imagination*. A comprehensive introduction to VSD, a field that Friedman has pioneered and developed. Contains practical instructions for organizing VSD.
- Marc Steen and Ibo van de Poel (2012): **Making values explicit during the design process,** *IEEE Technology and Society Magazine*, 31 (4), 63–72. One of my first attempts at VSD; a conceptual case study about values at play in the TA2 project.
- Paul Hayes, Ibo van de Poel, and Marc Steen (2020): **Algorithms and values in justice and security,** *AI & Society*, 35, 533–555 (open access). We discuss a range of relevant values and various interdependencies and tensions between these values.
- More about *Capability Driven Design*, by Annemarie Mink, is online: design4wellbeing.info. More about *Capability Sensitive Design*, in this article: Naomi Jacobs (2020): **Capability Sensitive Design for health and wellbeing technologies,** *Science and Engineering Ethics*, 26, 3363–3391 (open access), available online: https://doi.org/10.1007/s11948-020-00275-5
- You can find a *Societal and Ethical Impact Canvas* online (ethicsforpeoplewhoworkintech.com); it is meant to help your team to go from jointly clarifying desired impacts, to discussing human capabilities that are relevant for the project, and ways to enable people to develop these capabilities, to issues related to creating and maintaining an innovation ecosystem.

Chapter 19. Responsible Innovation

- Jack Stilgoe, Richard Owen, and Phil Macnaghten (2013): **Developing a framework for responsible innovation,** *Research Policy*, 42, 1568–1580. A relatively accessible introduction to Responsible Innovation; they introduced the four key dimensions that I discuss.

- Marc Steen and Joram Nauta (2020): **Advantages and disadvantages of societal engagement**: A case study in a research and technology organization, *Journal of Responsible Innovation*, 7, 3, 598–619 (open access). Discusses different methods for organizing societal engagement and pros and cons of societal engagement; it also contains the case study of making the Strategy Advisory Councils more inclusive and diverse.
- More about **Technology assessment**, https://en.wikipedia.org/wiki/Technology_assessment; also Participatory Technology Assessment (involving social actors) and Constructive Technology Assessment (steering the construction of technology). Currently, a term like ELSA (Ethical, Legal, Social Aspects) is used to refer to similar concerns and efforts.
- Ben Schneiderman (2022): *Human-centred AI*. Chapter 8 presents a two-dimensional framework, with *computer automation* and *human control* as two axes (rather than as opposing ends on one axis). Schneiderman advocates striking a balance between *automation* and *control* and avoiding too much and too little of each. I borrowed these concepts.
- Marc Steen, Tjerk Timan, and Ibo van de Poel (2021): **Responsible innovation, anticipation and responsiveness**: Case studies of algorithms in decision support in justice and security, and an exploration of potential, unintended, undesirable, higher-order effects, *AI and Ethics*, 1 (4), 501–515 (open access). We organized this articled around the four principles of the *High Level Expert Group*: human autonomy, prevention of harm, fairness, and explicability.
- Marc Steen (2021): **Slow innovation**: The need for reflexivity in Responsible Innovation, *Journal of Responsible Innovation*, 8 (2), 254–260 (open access). In this essay, I explore 'an alternative to the putative need for speed, the primacy of efficiency and the assumption that faster is better'.
- John Thackara (1999): *An unusual expedition* (Preface), in *Presence: New media for older people*, Kay Hofmeester and Esther De Charon de Saint Germain, editors. The quote about meeting 'putative users' is from pages 8–9.
- Marc Steen, Martin Sand, and Ibo Van de Poel (2021): **Virtue ethics for Responsible Innovation**, *Business and*

Professional Ethics Journal, 40 (2), 243–268 (open access). We discuss a range of virtues that are associated with responsibility and with innovation, and present reflexivity (practical wisdom, *phronesis*) as a key virtue.

- Chris McPhee, Martin Bliemel, and Mieke van der Bijl-Brouwer (2018): **Editorial: Transdisciplinary Innovation,** *Technology Innovation Management Review*, 8 (8), 3–6. They provide a very accessible introduction to Transdisciplinary Innovation (and to a special issue on that topic).

Chapter 20. What does your next project look like?

Additional online resources are available at **ethicsforpeoplewhoworkintech.com**. If you want to learn more about ethics in relation, specifically, to *Artificial Intelligence*, I can recommend these books and report:

- Marc Coeckelbergh (2020): *AI ethics.* Excellent introduction that goes into several key questions that are pertinent to AI. Also, I appreciated Coeckelbergh's discussion of the climate crisis and how it relates to the design and application of advanced technologies.
- Virginia Dignum (2019): *Responsible artificial intelligence: How to develop and use AI in a responsible way.* Excellent introduction. I have found the distinctions that Dignum makes between 'ethics in design', 'ethics by design', and 'ethics for design(ers)' particularly useful.
- John Havens (2016): *Heartificial intelligence: Embracing our humanity to maximize machines.* Very accessible. What I especially like about this book is that each chapter starts with fiction. This inspired me to write fiction (Chapters 8, 10, 12, 14).
- David Leslie and others (2021): *Artificial intelligence, human rights, democracy and the rule of law: A primer.* The Alan Turing Institute and the Council of Europe. Available online, https://www.turing.ac.uk/research/publications/ai-human-rights-democracy-and-rule-law-primer-prepared-council-europe

Gratitude

I would like to express my gratitude to various people who, in diverse ways, enabled and supported my work. In no particular order, I would like to thank: Hugo Letiche and Jan Buijs († 2015), who supervised my PhD research, and fellow students of the PhD program of the University for Humanistic Studies; co-workers and former co-workers Suzanne Ogier, Joram Nauta, Reijer Gaasterland, Jan-Willem Streefkerk, Helma van den Berg, Mark Neerincx, Tjerk Timan, Jurriaan van Diggelen, Freek Bomhof, Eric Veldkamp, Sabine Veenstra, and others who have collaborated in promoting societal engagement, ethical deliberation, and diversity and inclusion in various projects; project managers who made room for such efforts, like Doug Williams (TA2) and Sharon Prins (WeCare); people in academia for their research and insightful publications, notably Ibo van de Poel (also for our fruitful collaboration), Lambèr Royakkers (also for comments on the manuscript), Philip Brey, Mark Coeckelbergh, and Peter Paul Verbeek; and the many people whom I have worked with, in large and small companies, and in Dutch ministries and government agencies. The real action happens when we develop, apply, and evaluate methods and technologies in practice. I would also like to thank Emily Riederer for her review of the manuscript and Mina Witteman for her advice for the four fiction scripts. Of course, any remaining errors are entirely mine. Lastly, I wish to express gratitude for my son, Koen, for reminding me of the need to present complex matters in manageable forms, and for my wife Cynthia, for her support and for reminding me of worthwhile matters in life, other than reading and writing.

About the author

Marc Steen lives and works in the Netherlands. He received MSc, PDEng, and PhD degrees in Industrial Design Engineering at Delft University of Technology, which has pioneered human-centred design, transdisciplinary innovation, and innovation management. After working at Philips Consumer Electronics and KPN Research, he joined TNO (The Netherlands Organisation for applied scientific research, an independent research and technology organization) where he has worked in research, design, consultancy, and project management.

Over the years, Marc has become increasingly interested in the role of technology and innovation in people's daily lives and in society at large. He participated in a part-time PhD programme at the University of Humanistic Studies in Utrecht, where he developed an interest in the ethics involved in the design and application of technology. Additionally, he has participated in diverse courses and training programs, notably in marketing strategy, project management, coaching of professionals, and law. His mission is to support organizations to use technologies in ways that help to create a just society and promote people's flourishing. He asks uneasy questions about technologies, especially about the ethics involved in the design and application of algorithms and artificial intelligence. More information at marcsteen.nl.

Index

Printed in the United States
by Baker & Taylor Publisher Services